SpringerBriefs in Ecology

For further volumes:
http://www.springer.com/series/10157

O.W. Van Auken • J.K. Bush

Invasion of Woody Legumes

 Springer

O.W. Van Auken
The Department of Biology
University of Texas at San Antonio
San Antonio, TX, USA

J.K. Bush
Department of Earth and Env. Sciences
University of Texas at San Antonio
San Antonio, TX, USA

ISSN 2192-4759   ISSN 2192-4767 (electronic)
ISBN 978-1-4614-7198-1   ISBN 978-1-4614-7199-8 (eBook)
DOI 10.1007/978-1-4614-7199-8
Springer New York Heidelberg Dordrecht London

Library of Congress Control Number: 2013938376

Printed on acid-free paper

Springer is part of Springer Science+Business Media (www.springer.com)

Preface

No, this is not a science fiction book about space aliens and their attempts to take over Earth. It is about native woody legumes and their invasion or more correctly their encroachment into arid and semiarid C_4 tropical and subtropical grasslands. Invasion is the entry and establishment of a species in a new community, but the species that is establishing is from afar, usually another continent. Encroachment is the entry and establishment of a species in a new community, but the species that is establishing is from nearby, perhaps an adjacent community, or the species has increased in density locally. These changes usually are noted in grasslands, typically it is encroachment, and it is occurring all over the world. Generally the woody legumes are involved in the warmer grasslands.

Why is it important to understand this woody plant encroachment phenomenon? Because arid and semiarid drylands cover approximately 41 % of the terrestrial surface of the Earth. In addition, approximately 2.4 billion people live in these dryland habitats. The herbaceous productivity of many dryland or arid zone grasslands all over the world has decreased, resulting in a decrease of domestic animal production in the same grasslands. The herbaceous productivity has decreased because of the increased density and productivity of woody plants in these areas.

Prosopis and *Acacia* are genera of leguminous woody shrubs and small trees that have increased in density and mass in many of these arid and semiarid C_4, temperate, subtropical, and tropical grasslands. The systematics of the genus *Acacia* has recently been changed, and these changes will be addressed. In addition, this book will focus mostly on a few species of the former genus *Acacia* and *Prosopis* mostly on those present in the North American southwest. The woody legumes ability to grow, spread, invade, or encroach into arid and semiarid grasslands has not been easily explained and seems to be dependent on a number of factors that will be examined. Water and nutrients requirements, competitive abilities, successional status, global change, and management phenomenon, as well as ungulate herbivory will be discussed. The importance of past, present, and future grassland fires and potential restoration will also be included.

Understanding the biological and ecological characteristics of the woody legumes and their competitors and reasons for the woody legumes ability to invade or encroach in these warm season grasslands is the major goal and focus of this book. In searching the literature, there were approximately 10,000 citations found when the main genera of woody legumes were searched over the past 10 years. We will not include all of these research citations, but we will try to include the most important papers found, especially ecological ones.

We have been interested in the woody legumes for many years, but did not understand their ecological position or importance in a community for a long time. Understanding their ecological position and community importance has been the driving force of our research. We have learned a lot, but there are still many unanswered questions. This book represents a considerable amount of research that we and others have done on the woody legumes over many years, and there are still many interesting questions to ask and answer.

We thank Kelly Jo Stephens for the considerable help in formatting this publication and helping keep everything on track and focused. Stephanie Elliott answered many questions about the figures and photographs that were included. There are too many former students, currently faculty and friends to name, but thanks to all for the help, support, and encouragement.
January 2, 2013

San Antonio, TX, USA   O.W. Van Auken
                                                                                      J.K. Bush

Contents

Contents

Chapter 1
Introduction

Prosopis and *Acacia* are two genera of woody legumes that occur worldwide, mostly in dry lands in the arid zone and in the arid and semiarid, temperate, subtropical, and tropical grasslands, savannas, and woodlands (Figs. 1.1 and 1.2) (Simpson 1977; Heywood 1978; New 1984; Tame 1992). Some *Acacia s. l.* (sensu lato) species in eastern Africa seem to be riparian species, and a few Australia species seem to be successional species and appear be a part of forest communities which are in wetter areas. In addition, the ecological position of a number of New World tropical *Acacia* species *s. l.* is not clear. Many of these New World tropical species seem to be found in disturbed areas (Janzen 1974), but these species and communities will not be described or discussed. In this book, we will mostly deal with a few members of the genus *Acacia* or more specifically with some members of the former genus *Acacia s. l.*, but there will be some general information about *Prosopis* and references to *Prosopis* when comparisons are important and justified. Recently, the genus *Acacia s. l.* has been divided into five genera including *Senegalia* and *Vachellia* (Maslin et al. 2003; CPBD 2004; Seigler and Ebinger 2005; Maslin 2006; Seigler et al. 2007; Crow and Ritter 2012). We will focus on the ecology of a handful of specific members of the genera *Senegalia* and *Vachellia* with a few references to *Prosopis*. Most of the species of the new genus *Acacia* are found in Australia, and these species and communities will not be discussed here.

Members of these genera have been used by native and modern people for construction, tanning hides, medicine, dyes, soaps, perfumes, fibers, animal fodder, and fuel (Simpson 1977; New 1984; Tame 1992; Dharani 2006). Some species of *Senegalia* and *Vachellia* have also been used in soil conservation and to some extent in erosion control. However, one of the most important characteristics of many of these species is that they are legumes, are associated with nitrogen-fixing bacteria, and add substantial amounts of biologically fixed nitrogen to the soil, especially degraded soils or low nitrogen soils (Simpson 1977; West and Klemmedson 1978; Tame 1992; Bush 2008; Khan et al. 2010; Tye and Drake 2012).

O.W. Van Auken and J.K. Bush, *Invasion of Woody Legumes*, SpringerBriefs in Ecology 4, DOI 10.1007/978-1-4614-7199-8_1, © Springer Science+Business Media New York 2013

Fig. 1.1 Approximate worldwide distribution of *Prosopis* with the majority of species (35) in South America (modified from Simpson 1977)

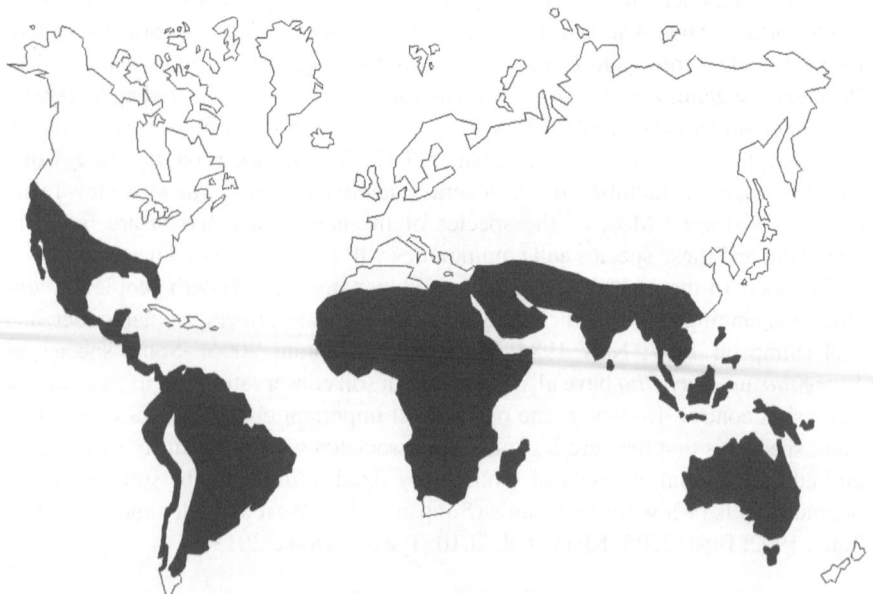

Fig. 1.2 Approximate worldwide distribution of *Acacia* (*s. l.*) (modified from New 1984)

These woody legumes are important components of many worldwide arid and semiarid ecosystems (Whittaker 1975; Crisp and Lange 1976; Simpson 1977; Benson 1979; Felker 1981; Barbour and Billings 1988). Today there are reports of large expanses of *Prosopis, Senegalia,* and *Vachellia* shrub lands in southwestern North America including large areas in the South Texas Plains and the plains of northern Mexico (Gould 1969; Correll and Johnston 1979). In the past this region or at least large parts of it were described as grassland or savanna (Johnston 1963; Gould 1969; Correll and Johnston 1979; Anderson 2006). Other studies including interpretations from the earliest travelers (late 1600s) suggest that this area was a mosaic of vegetation (see Inglis 1962) with intermittent thickets or motts, probably consisting of legumes like *Prosopis* or various *Senegalia* and *Vachellia* species, probably on deeper soils along some of the creeks or drainages or on rocky outcrops protected from fire. Some of the earliest reports of the vegetation were from travelers in expeditions through northeastern Mexico through the now thorn woodlands (Tamaulipan woodlands and south Texas Plains) north through the San Antonio area of Texas. The larger and permanent streams and rivers probably had riparian or gallery forests as they do today (Bush and Van Auken 1984). Much of the uplands were most likely savannas with grassland and shrub land phases, but the grasslands have apparently been shrinking with increased invasion or encroachment of woody plants as domestic grazing increased and fire frequency has been reduced. This phenomenon of decreasing grasslands with a concomitant increase of woodlands, savannas, or shrub lands is occurring or has occurred worldwide (Van Auken 2000, 2009; Knapp et al. 2008; Scholes 2009; Wigley et al. 2010; Sharp et al. 2012).

Invasion is really the establishment, growth, and survival of a species in a new area or community with the species coming from afar, probably another continent (Mooney and Drake 1991). Encroachment is the increase in density and spread of native species into closely associated communities, usually grasslands. These species have been present in these same or adjacent communities for thousands of years (Inglis 1962). Encroachment is the phenomenon occurring in the American southwest and in other grasslands all over the world (Van Auken 2000, 2009). The process of encroachment is usually considered as the entry of woody plants, in many cases woody legumes, into grasslands. There have been many studies of the changes wrought by these woody plants as they enter grasslands, especially as they affect grass forage and domestic animal production (Browning and Archer 2011; Allred et al. 2012). However, the process appears to be very similar or the same as early secondary succession (Begon et al. 2006) which will be presented later.

Most of the former grasslands of this area of southwestern North America today are covered with various woody species including *Prosopis glandulosa* (mesquite or honey mesquite and possibly other species of *Prosopis*), *Celtis pallida* (desert hackberry), *Leucophyllum frutescens* (Texas sage), *Aloysia gratissima* (white brush), *Baccharis neglecta* (baccharis), and other woody species as well as various woody legumes including *Vachellia farnesiana* (=*Acacia farnesiana*=*A. smallii,* huisache), *V. rigidula* (=*A. rigidula,* blackbrush), *V. tortuosa* (=*A. tortuosa,* huisachillo), *V. neovernicosa* (=*A. neovernicosa,* viscid acacia), *V. constricta* (=*A. constricta,* mescat acacia), *Senegalia berlandieri* (=*A. berlandieri,* guajillo, or fern acacia), *S. greggii* (=*A. greggii,* catclaw acacia), *S. roemeriana* (=*A. roeme-riana,* Romer's acacia), and others (Correll and Johnston 1979; Van Auken and Bush 1985; Jurena and Van Auken 1998; Powell 1998; Seigler and Ebinger 2005; Seigler et al. 2007).

There are three species of *Prosopis* in southwestern parts of Asia and all across central Africa (Fig. 1.1) (Simpson 1977), with 35 species in South America and nine species from North America. The two major species of *Prosopis* in North America (*P. velutina* and *P. glandulosa*) occur mostly in the arid and semiarid regions of the southwestern United States and northern parts of Mexico (Fig. 1.1). *Prosopis velu-tina* is present in southeastern California, southern Arizona, and northwestern Mexico. *Prosopis glandulosa* is more widespread and is reported from southern Kansas, southwestern Oklahoma, central and western Texas, northern Mexico, eastern and southern New Mexico, eastern and southern Arizona, southwestern California, and Baja California (Correll and Johnston 1970; Simpson 1977).

We originally planned to include the old genus *Acacia s. l.* in this book, and we will include it briefly in the next chapter, to put our work in prospective in relation to the former and current genus *Acacia s. l.* The genera with species that we have worked with are *Senegalia* and *Vachellia* which, for the most part, we will concen-trate on here. However, *Prosopis* will be included as appropriate.

Members of the old genus *Acacia* (*s. l.*) are found in the same general area as *Prosopis* but more widespread, mostly temperate, tropical and subtropical, arid and semiarid C_4 grasslands, savannas, and woodlands in both the Old and New World (Fig. 1.2). However, *Prosopis* is not reported from southeast Asia or Australia. The *Acacias* are usually found below 1,000 m, but some have been reported up to 2,100 m (Dharani 2006).

In the following chapters we will suggest that the invasion or encroachment of woody legumes into grasslands is the start of secondary succession in these grasslands (Begon et al. 2006). We propose to present and discuss why we think this is the case. We will also show the structure of some of these savanna and woodland communities and demonstrate the factors that seem to control or determine the growth of the main species in these communities. In addition, we will examine the importance of competition between the grasses and the legumes and the distribution of species in the communities. Other topics of importance that will be examined include global climate change and how management of these communities should commence. Are these grasslands, woody legume savannas, and woodlands ecologically stable communities? Could they be restored or should they be restored is also a topic. Then, we will pull all of these fac-tors together in a discussion and projection of what we think will happen in the future.

References

Allred BW, Fuhlendorf SD, Smeins FE, Taylor CA (2012) Herbivore species and grazing intensity regulate community composition and an encroaching woody plant in semi-arid rangeland. Basic Appl Ecol 13:149–158

Anderson RC (2006) Evolution and origin of the Central Grasslands of North America: climate, fire and mammalian grazers. J Torrey Bot Soc 133:626–647

Barbour MG, Billings WD (1988) North American terrestrial vegetation. Cambridge University Press, New York

Begon M, Townsend CR, Harper JL (2006) Ecology: from individuals to ecosystems, 4th edn. Blackwell, Malden, MA

Benson L (1979) Plant classification. Heath, Lexington, MA

Browning DM, Archer S (2011) Protection from livestock fails to deter shrub proliferation in a desert landscape with a history of heavy grazing. Ecol Appl 21:1629–1642

Bush JK (2008) Soil nitrogen and carbon after twenty years of riparian forest development. Soil Sci Soc Am J 72:815–822

Bush JK, Van Auken OW (1984) Woody species composition of the upper San Antonio River gallery forest. Tex J Sci 36:139–145

Correll DS, Johnston MC (1970) Manual of the vascular plants of Texas. Texas Research Foundation, Renner

CPBD (2004) The name Acacia retained for Australian species, 1–3. Center for Plant Biodiversity Research

Crisp MD, Lange RT (1976) Age structure, distribution and survival under grazing of the arid zone shrub. Acacia burkittii Oikos 27:86–92

Crow T, Ritter M (2012) Changes to the Botanical Code and what they mean for western North American Botany. Madrono 59:169–170

Dharani N (2006) Field guide to the Acacias of east Africa. Struik, Cape Town

Felker P (1981) Use of tree legumes in semiarid regions. Econ Bot 35:174–186

Gould FW (1969) Texas plants: a checklist and ecological summary. Texas Agriculture Experiment Station Bull, College Station

Heywood VH (1978) Flowering plants of the world. Oxford University Press, London, 335

Inglis JM (1962) A history of vegetation on the Rio Grande Plain. Texas Parks and Wildlife Department. Austin, Texas

Janzen DH (1974) Swollen-thorn acacias of Central America. Smithsonian Contrib Bot 15:1–140

Johnston MC (1963) Past and present grasslands of southern Texas and northeastern Mexico. Ecology 44:456–466

Jurena PN, Van Auken OW (1998) Woody plant recruitment under canopies of two acacias in a southwestern Texas shrubland. Southwest Nat 43:195–203

Khan B, Ablimis A, Mahmood R, Qasim M (2010) *Robinia Pseudoacacia* leaves improve soil physical and chemical properties. J Arid Lands 2:266–271

Knapp AK, Briggs JM, Collins SL, Archer SR (2008) Shrub encroachment in North American grasslands: shifts in growth form dominance rapidly alters control of ecosystem carbon inputs. Glob Chang Biol 14:615–623

Maslin BR (2006) Generic and infrageneric names in *Acacia* following retypification of the genus. World Wide Wattle 1–3

Maslin BR, Miller J, Seigler DS (2003) Overview of the generic status of *Acacia* (Leguminosae: Mimosoideae). Aust Syst Bot 16:1–18

Mooney HA, Drake JA (1991) Ecology of biological invasions of North America and Hawaii. Springer, New York

New TR (1984) A biology of Acacias. Oxford University Press, Melbourne

Powell AM (1998) Trees and shrubs of the Trans-Pecos and adjacent areas. University of Texas Press, Austin

Scholes RJ (2009) Syndromes of dryland degradation in southern Africa. Afr J Range Forage Sci 26:113–125

Seigler DS, Ebinger JE (2005) New combinations in the genus *Vachellia* (Fabaceae:Mimosoideae) from the New World. Phytologia 87:139–171

Seigler DS, Ebinger JE, Kerber A (2007) Characterization of thorn-scrub woodland communities at the Chaparral Wildlife Management Area in the south Texas Plains, Dimmit and La Salle Counties, Texas. Phytologia 89:241–257

Sharp EA, Spooner PG, Millar J, Briggs SV (2012) Can't see the grass for the trees? Community values and perception of tree and shrub encroachment. Landscape Urban Plann 104:260–269

Simpson BB (1977) Mesquite: its biology in two desert ecosystems. Dowden Hutchinson & Ross Inc, Stroudsburg, PA

Tame T (1992) Acacias of Southeast Australia. Kangaroo, Kenthurst, Australia

Tye DRC, Drake DC (2012) An exotic Australian *Acacia* fixes more N than a coexisting indigenous *Acacia* in a south African riparian zone. Plant Ecol 213:251–257

Van Auken OW (2000) Shrub invasions of North American semiarid grasslands. Annu Rev Ecol Syst 31:197–215

Van Auken OW (2009) Causes and consequences of woody plant encroachment into western North American Grasslands. J Environ Manage 90:2931–2942

Van Auken OW, Bush JK (1985) Secondary succession on terraces of the San Antonio River. Bull Torrey Bot Club 112:158–166

West NE, Klemmedson JO (1978) Structural distribution of nitrogen in desert ecosystems. In: West NE, Skujins JJ (eds) Nitrogen in desert ecosystems. Dowden, Hutchinson and Ross, Inc, Stroudsburg, PA, pp 1–17

Whittaker RH (1975) Communities and ecosystems. MacMillan Publishing Co, New York, 162

Wigley BJ, Bond WJ, Hoffman MT (2010) Thicket expansion in a South African savanna under divergent land use: local vs. global drivers? Glob Chang Biol 16:964–976

Chapter 2
Species Systematics

Difficulties with the systematics or taxonomy of the genus *Acacia s. l.* are not new. Problems with the systematics of the swollen-thorn acacias were reported 39 years ago (Janzen 1974). This difficulty has continued up to the present day. *Acacia s. l.* was a very large genus 25–50 years ago with 600–1,000 species estimated to occur worldwide (Turner 1959; Janzen 1974; New 1984). More recently, about 1,350 species have been reported (CPBD 2004; Dharani 2006), including 954 from Australia, 230 in the Americas, 132 in Africa, 18 in Asia, and possibly ten from some of the Pacific Islands. There is probably some overlap of species occurrence, counts, and descriptions because of population and clinal variation (Janzen 1974; New 1984). However, a number of studies have examined the relationships between the many species of *Acacia s. l.* and suggested taxonomic modifications of the genus because it is polyphyletic (see New 1984; Seigler et al. 2006). Both morphologic, genetic, and molecular evidence suggested that the former genus *Acacia s. l.* was polyphyletic and should be separated into as many as five genera including *Acaciella, Mariosousa, Senegalia,* and *Vachellia,* with the retention of *Acacia* (Seigler and Ebinger 2005; Maslin 2006; Seigler et al. 2007; Crow and Ritter 2012). The genera *Senegalia* and *Vachellia* include many shrubs and small trees that are common in arid and semiarid parts of the American southwest. These genera include species that we are familiar with and species that we have worked with which, for the most part, we will include and focus on here.

In the new genera *Senegalia* and *Vachellia,* there appear to be 146 taxa in the Americas with a total of 365 taxa worldwide (Seigler and Ebinger 2005; Seigler et al. 2006). There seem to be about 86 species and two varieties of *Senegalia* in the Americas, 69 taxa in Africa, 43 in Asia, and 2 in Australia with 8 species overlapping in areas. This would be a total of 202 taxa worldwide (Seigler et al. 2006). For *Vachellia,* there appear to be 60 taxa in the New World, 73 in Africa, 36 in Asia, and 9 in Australia with 15 overlapping for a total of 163 taxa worldwide (Seigler and Ebinger 2005).

Fig. 2.1 *Left* picture above shows the presence of prickles on the stem of *Senegalia greggii* (=*Acacia greggii*, catclaw acacia). On the *right* is a picture of *Vachellia farnesiana* (=*Acacia farnesiana*, huisache) that has two stipular spines and usually the absence of prickles. Photographs are courtesy of Ms. Wendy Leonard and were taken in the San Antonio, Texas area

Many of the Old and New World acacias seem related. For example, the African swollen-thorn acacias seem to have an ant mutualism similar to the Central American swollen-thorn acacias (Janzen 1974). *Vachellia farnesiana* (=*Acacia farnesiana* =*A. smallii*) is a widespread species in the New World, occurring across the southeastern USA, through most of southeastern, south central, and southwestern Texas. Its distribution continues south and west through the arid and semiarid regions of northern Mexico, southern New Mexico, Arizona, and California and south into the tropical region including the Caribbean, central America, and northern South America (Turner 1959; Correll and Johnston 1979; Turner et al. 2003). In addition, it is reported from parts of southern Africa, arid parts of Australia, and some Pacific Islands (Dharani 2006). It seems to be a New World species, possibly transplanted in recent times into various Old World locations. Apparently, approximately 12 % of the species in the former genus *Acacia s. l.* mostly species in the genera *Senegalia* and *Vachellia* are reported from the Americas (Seigler and Ebinger 2005; Seigler et al. 2006), and 12 or 13 species are reported in Texas and surrounding areas (Turner 1959; Correll and Johnston 1979; Turner et al. 2003; Seigler et al. 2007).

Descriptions of the morphological characteristics that separate the genera *Senegalia* and *Vachellia* can be found in Seigler's recent publications (Seigler and Ebinger 2005; Seigler et al. 2006). Characteristics in *Senegalia* would include the lack of stipular spines, the presence of vegetative stipules, and the occurrence of prickles and a few other characteristics (Fig. 2.1). On the other hand, *Vachellia* has stipular spines (usually two) with the absence of prickles (Fig. 2.1). In addition, they list the names of the species of *Senegalia* and *Vachellia* present in the Americas. They also present some molecular genetics including a most parsimonious consensus tree comparing a number of *Senegalia* and *Vachellia* species, with a few *Acacia* and other species as out groups. They show that all of the *Senegalia* species are related and are in one group; in addition, all of the species of *Vachellia* are related and are in another group. These two groups of species are all separated from all of the other species used to generate the tree.

References

Correll DS, Johnston MC (1979) Manual of the vascular plants of Texas. The University of Texas at Dallas, Richardson, TX

CPBD (2004) The name Acacia retained for Australian species, 1–3. Center for Plant Biodiversity Research

Crow T, Ritter M (2012) Changes to the botanical code and what they mean for western North American Botany. Madrono 59:169–170

Dharani N (2006) Field guide to the Acacias of East Africa. Struik, Cape Town

Janzen DH (1974) Swollen-thorn *Acacias* of Central America. Smithsonian Contrib Bot 15:1–140

Maslin BR (2006) Generic and infrageneric names in *Acacia* following retypification of the genus. World Wide Wattle, 1–3

New TR (1984) A biology of Acacias. Oxford University Press, Melbourne

Seigler DS, Ebinger JE (2005) New combinations in the genus *Vachellia* (Fabaceae:Mimosoideae) from the New World. Phytologia 87:139–171

Seigler DS, Ebinger JE, Miller JT (2006) New combinations in the genus *Senegalia* (Fabaceae:Mimosoideae) from the New World. Phytologia 88:38–93

Seigler DS, Ebinger JE, Kerber A (2007) Characterization of thorn-scrub woodland communities at the Chaparral Wildlife Management Area in the south Texas Plains, Dimmit and La Salle Counties, Texas. Phytologia 89:241–257

Turner BL (1959) The legumes of Texas. University of Texas Press, Austin

Turner BL, Nichols H, Denny G, Doron O (2003) Atlas of vascular plants of Texas. Brit, Fort Worth, pp 648

Chapter 3
Encroachment and Secondary Succession

There are estimations suggesting that approximately 60 million ha of grasslands in the American southwest have been encroached by various woody species and converted to savannas, shrublands, or woodlands (Humphrey 1958; Grover and Musick 1990). Interestingly, this is more than the total estimated area of these grasslands (Laurenroth 1979). Thus, the composition, cover, density, productivity, and structure of most if not all of these grasslands has changed. In addition, it has been suggested that 220–330 million ha of all North American grasslands have been encroached by woody species (Knapp et al. 2008). Not all of this woody plant encroachment is by woody legumes, but a considerable amount is (Van Auken 2009). *Prosopis, Senegalia,* and *Vachellia* are genera of woody legumes with major encroaching species in southwestern North American grasslands, but they are not the only species.

Encroachment of grasslands by woody plants from adjacent communities seems to be the start of a process called succession. Succession or ecological succession is the "non-seasonal, directional and continuous pattern of colonization and extinction on a site by a species populations" (Begon et al. 2006; Smith and Smith 2012). There are really two types of terrestrial succession recognized by ecologists. Primary succession is found or reported on newly exposed areas or landforms. The area in question has not been previously occupied by biotic species or communities of any type, and there is no evidence of any soil development. Secondary succession on the other hand occurs on a site when a community has been removed partially or completely, and for the most part, only the soil remains. The causes of the community and species removal are various factors. It could be fires, hurricanes, tornados, high winds, and disease of some type or anthropogenic activities including tree harvest, agriculture, or grazing. The start of recovery of the community in terms of secondary succession begins after the disturbance but seems to be called encroachment in many cases. This is when the woody plants start to establish in the disturbed grassland community. The return to the previous pre-disturbance community can take a short time or a very long time depending on the type of community being replaced and the environmental conditions of the area together with or especially climatic conditions.

O.W. Van Auken and J.K. Bush, *Invasion of Woody Legumes*, SpringerBriefs in Ecology 4, DOI 10.1007/978-1-4614-7199-8_3, © Springer Science+Business Media New York 2013

Ecological succession has been studied for many years. In 1863, Henry David Thoreau recognized that upland pine communities were replaced after logging by hardwood communities (see Spurr and Barnes 1973). Clements (1916) suggested that succession would proceed to a climax community that would dominate a particular climatic region. This appears to be too simplistic an approach for a complex theory. The climate in a local area may control the climax plant community type, but usually there is a combination of factors that control the type of climax community in an area in addition to climate, such as soil conditions, topography, fire, and possibly others. All communities appear to continue to change in time but late in succession as the climax community is approached, the rate of change slows. Disturbances, large and small, would result in patches or gaps in a community and seem to reset a part of the community to an earlier stage of succession, which could slow the successional process or the approach to the climax community (Begon et al. 2006; Smith and Smith 2012).

Many of the woody legume communities have seemed difficult to study because conditions necessary to understand the start of the succession have not been well recognized or described. This may be in part because of the relatively long time interval required to start the process in relatively dry or arid conditions and areas. Or, it may be the relatively long time it takes for the woody legumes to overtop the grasses and be recognized as encroaching. Or, it could be the level or degree of disturbance, which may be continuous or discontinuous. The disturbances could also interact in various ways with a highly variable climatic condition such as amount of rainfall. Curiously, for the most part, connections have not apparently been made between the encroachment process and secondary succession. This incursion or encroachment of woody legumes into grasslands has been going on for a long time, but has not been recognized by many as secondary succession.

There are many different ways to look at and explain succession (Begon et al. 2006; Smith and Smith 2012). But the driving force or forces of succession are much harder and more difficult to find, describe, and explain. These are the biological mechanisms that underlie succession. There are lots of suggestions concerning mechanisms, and there are many mechanisms that fit certain community types and conditions. One theory that seems to make considerable sense concerning woody legume encroachment and the successional sequences that follow is the resource ratio hypothesis (Tilman 1985). This theory suggests that the dominance of a species in time and space during terrestrial succession is dependent on the relative amount or availability of two potentially limiting resources. The two resources that he focused on were light levels and a limiting soil resource such as nitrogen (Fig. 3.1).

Surface light levels are high early in terrestrial succession (Fig. 3.1) and then decrease as shade-tolerant woody species other than the woody legumes establish, grow, and become a dominant part of the overstory or canopy. The second resource, soil nitrogen level, is low early in succession and then increases late in succession. In the case of woody legumes in encroachment or secondary succession, the sequence is slightly different. Soil nitrogen levels are low early in succession, but the woody legumes or more particularly their symbiotic nodule-forming bacteria fix atmospheric nitrogen. This nitrogen is used by the legume, and then the nitrogen is

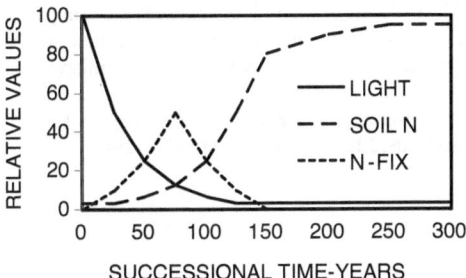

Fig. 3.1 Theoretical figure showing relative changes in surface light levels, nitrogen fixation, and soil nitrogen levels during secondary succession in a warm season C₄ arid zone grassland. Early secondary succession or encroachment would be on the *left*, development of the legume savanna or woodland would be toward the *center*, and the late successional nonlegume woodland or forest would be on the *right*

added to the soil via leaf and litter fall, followed by decomposition by soil bacteria and fungi. The nitrogen is released from the litter, and soil nitrogen levels increase. Next, this nitrogen is available to nonlegumes for their growth and metabolism. As a result, other woody species requiring higher levels of soil nitrogen but are tolerant of shade establish in the low light environment below the legume canopy. These nonlegumes that are tolerant of low light conditions then overgrow and shade the high light requiring early successional, woody legumes that previously encroached the area (Van Auken 2000).

This is the encroachment process and successional sequence that has occurred and seems to be occurring in many southwestern North American warm season grasslands and probably other grasslands around the world. We will present evidence and results of experiments to show that this sequence is in fact true.

References

Begon M, Townsend CR, Harper JL (2006) Ecology: from individuals to ecosystems, 4th edn. Blackwell, Malden, MA

Clements FE (1916) Plant succession: analysis of the development of vegetation. Carnegie Institute of Washington Publication No. 242, Washington, DC

Grover HD, Musick HB (1990) Shrubland encroachment in southern New Mexico, USA: an analysis of desertification processes in the American southwest. Climate Change 17:305–330

Humphrey RR (1958) The desert grassland: a history of vegetation change and an analysis of causes. Bot Rev 24:193–252

Knapp AK, Briggs JM, Collins SL, Archer S (2008) Shrub encroachment in North American Grasslands: shifts in growth form dominance rapidly alters control of ecosystem carbon inputs. Glob Change Biol 14:615–623

Laurenroth WK (1979) Grassland primary production. In: French NR (ed) Perspectives in grassland ecology: results and applications of the US/IBP grassland biome study, 3–24. Springer, New York

Smith TM, Smith RL (2012) Elements of ecology, 8th edn. Benjamin Cummings, New York

Spurr SH, Barnes BV (1973) Forest ecology. The Ronald Press Company, New York
Tilman D (1985) The resource-ratio hypothesis of plant succession. Am Nat 125:827–852
Van Auken OW (2000) Shrub invasions of North American semiarid grasslands. Annu Rev Ecol
    Syst 31:197–215
Van Auken OW (2009) Causes and consequences of woody plant encroachment into western
    North American Grasslands. J Environ Manag 90:2931–2942

Chapter 4
Woody Legume Community Structure

Apparently many former grasslands and savannas have changed because of anthropogenic disturbances caused by the introduction of numerous domestic grazing animals (cattle and other species), the reduction of light, fluffy fuel (grasses by herbivory), and a concomitant reduction of fire frequency (Van Auken 2000, 2009). In southwestern North America, these changes have led to encroachment of various woody species including species of *Prosopis, Senegalia,* and *Vachellia* (the latter two previously acacias). The conditions required for the establishment (encroachment), growth, and dominance of the woody plants in many of these former grasslands have been difficult to understand. Global change phenomena including increased levels of atmospheric CO_2 and concomitant elevated temperature do not seem to be the main cause of the encroachment, which appears to be a management problem or phenomena (Van Auken 2000, 2009).

One of the most well-studied species of the former genus *Acacia s. l.* in the New World is *Vachellia farnesiana* (Fig. 4.1). It was reported on more that 1.1 million ha in south Texas alone with more than 20 % cover in many places (Smith and Rechenthin 1964) and probably 100 % cover in mid-successional, dense, woodland thickets or communities (Van Auken and Bush 1985; Bush et al. 2006). Although it is widespread and has a high density in places, there are still relatively few papers concerning its basic ecology (Van Auken and Bush 1985). Various management strategies have been used to try to control it because it is encroaching into many C_4 grasslands (Scifres 1980).

Prosopis (P. glandulosa, P. velutina, P. torreyana, or P. juliflora) another genus of woody legume has been reported on more that 38 million ha in the American southwest and northern Mexico (Van Auken 2000). Some arid or semiarid communities have relatively low *Prosopis* cover and density (Fig. 4.2) (Smith and Rechenthin 1964), but there are many apparent mid-successional, dense, woodland thickets where the cover approaches 100 % (Smith and Rechenthin 1964; Simpson 1977; Scifres 1980; Archer et al. 1988). It is a widespread genus and has a high density in places, and there is considerable concern about its increase in density in grasslands or former grasslands (Scifres 1980; Heitschmidt and Struth 1991).

Fig. 4.1 A relatively open *Vachellia farnesiana* savanna in south Texas is presented in the figure. The area is on the floodplain of the San Antonio River. The community is approximately 25-year post-encroachment or agricultural abandonment or 25 years into secondary succession. There are various C_4 grasses as well as herbaceous annual and perennial plants in the community. A few woody saplings of *Celtis laevigata* and *Prosopis glandulosa* are also present

Fig. 4.2 Photograph is a central Texas midgrass prairie demonstrating *Prosopis glandulosa* encroachment and relatively open early savanna community development with a relatively low woody plant density and cover

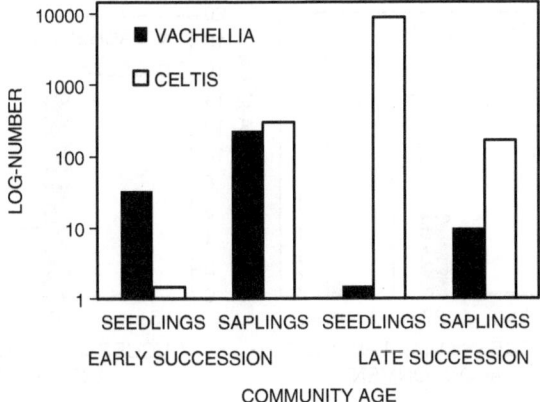

Fig. 4.3 Seedlings and saplings of *Vachellia farnesiana* and *Celtis laevigata* in early successional communities (5 years old) and mid-successional communities (33 years old) are compared. Note the scale is logarithmic (modified from Van Auken and Bush 1985). Using a *z*-statistic, all comparisons between species within a given age stand and within species between different age stands were significantly different for both seedling and sapling densities ($P < 0.05$)

However, its biology and basic ecology are still not fully understood (Archer et al. 1988; Van Auken and Bush 1989, 1997; Bush and Van Auken 1991, 1995; Archer 1994), although it has been extensively studied (see Browning and Archer 2011). Various management strategies have been used to try to control it because it has encroached into many C_4 grasslands, and there are a multitude of papers concerning the results of control strategies (Scifres 1980; Heitschmidt and Struth 1991; Taylor et al. 2012). Although *Prosopis* is worldwide in distribution and has been extensively studied, we will mostly compare it to *Senegalia* and *Vachellia*.

We will focus on the work done on *V. farnesiana* mainly from the southwestern part of the USA and northern Mexico. The chronological appearance and function of many of the woody legumes in these successional communities is still not completely understood. *Vachellia*, particularly *V. farnesiana,* and *Prosopis glandulosa* look like early successional species that establishes on low nutrient or low nitrogen soils (Van Auken et al. 1985; Archer et al. 1988) or in disturbances.

There are few studies that have demonstrated the presence of woody legumes like *V. farnesiana* as pioneer or early successional species (Van Auken and Bush 1985; Bush et al. 2006). The significance of high rates of biological nitrogen fixation in early successional communities and the role of woody legumes in these communities seems to be very important (Tilman 1985). Seedlings and saplings of *V. farnesiana* are found in early successional communities (disturbed or heavily grazed grasslands) but very few or not any in mid-successional communities (33 years old or older) (Fig. 4.3, note log scale).

A late successional and co-occurring species (*Celtis laevigata*) had very few seedlings but some saplings in the same early successional communities and a very large number of seedlings and some saplings in mid- or late successional communities

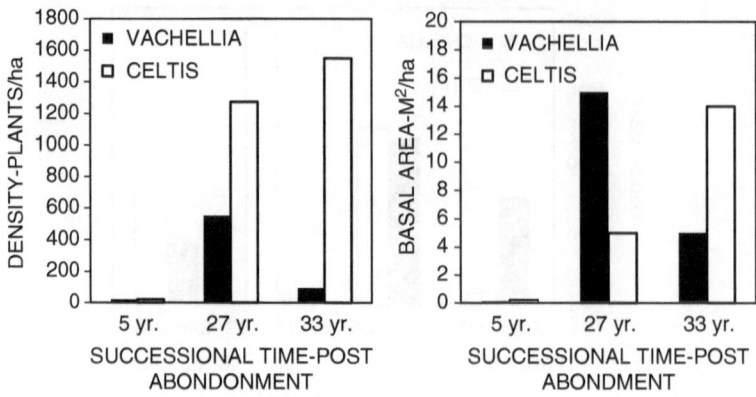

Fig. 4.4 Density (*left*) and basal area (*right*) of *Vachellia farnesiana* and *Celtis laevigata* in successional communities that are 5-year, 27-year, and 33-year postdisturbance (modified from Van Auken and Bush 1985). Standard error for basal area of *Vachellia farnesiana* in the 27-year community was 2.7 m²/ha, and for *Celtis laevigata* in 33-year community, it was 0.8 m²/ha

(Fig. 4.3) (Van Auken and Bush 1985). Woody legumes like *V. farnesiana* that appear to encroach into these arid and semiarid ecosystems occur at high density, high biomass, and high basal area early in succession (Fig. 4.4). The presence of these species and various community characteristics will change in time with the loss of the woody legume (Bush et al. 2006). *Vachellia farnesiana* is usually not found in mature woodland or forest communities as its density and basal area decline as succession proceeds and mature community dominants such as *C. laevigata* increase in density and basal area (Fig. 4.4) (Van Auken and Bush 1985; Bush and Van Auken 1989; Bush et al. 2006). *Celtis laevigata* and other mature community species increase their growth, including density and basal area, below the *Vachellia* canopy in the higher nitrogen soil and in the lower light environment.

Vachellia farnesiana appears to be growth suppressed in its own shade below its own canopy or by the shade of other canopy species (Bush and Van Auken 1986a; Lohstroh and Van Auken 1987). *Vachellia farnesiana* seedlings are found in open disturbed areas and open grassland or savanna associated with supposed parent trees (partial shade) early in succession, but they are not present when the canopy closes with community development probably because of low light levels below the canopy (Bush and Van Auken 1986a, b).

Soil levels of carbon, nitrogen, phosphate, calcium, and potassium below the *V. farnesiana* canopy were higher than values in gaps or intervening areas (Bush and Van Auken 1989) because of nitrogen fixation by the legume and subsequent litter fall and decomposition below the canopy. Soil resource levels in the intervening patches or gaps were lower than levels found below the *V. farnesiana* canopy. Similar trends are expected in other woody legume communities with similar rainfall, soil depth, and plant densities. However, in more arid savanna regions, the woody legume canopies may not coalesce or fuse, or it would take a longer time for this to happen and the communities would probably continue as woody legume savannas with higher levels of carbon and nitrogen below the canopy and lower levels in the intervening patches (Jurena and Van Auken 1998). This would probably be similar to what has been reported in other woody legume communities in the Tamaulipan region of south Texas, northeastern Mexico, and the Chihuahuan Desert in New Mexico (Archer et al. 1988; Schlesinger et al. 1996).

In time, the *Vachellia* or *Senegalia* species that are the focus of these patches of woodland in the savannas would die and disappear from these communities (Bush and Van Auken 1987a; Jurena and Van Auken 1998), similar to what happens to other woody legume patches or thickets in this region (Archer et al. 1988; Archer 1994). *Vachellia farnesiana* would not be able to replace itself in the savanna patches or woodland patches probably because of lower light levels below its canopy (Bush and Van Auken 1986a). Replacement species would be any number of trees or shrubs capable of starting growth in the higher nitrogen soil and the low light environment of the canopy understory (Bush and Van Auken 1986b, 1987b). The replacement species in relatively mesic areas, in parts of south Texas and northeastern Mexico, would probably be *Celtis laevigata* (hackberry or sugarberry), *Ulmus crassifolia* (cedar elm or fall elm), *C. pallida* (desert hackberry), *Bumelia lanuginosa* (bumelia), *Melia azedarach* (chinaberry), *Ehretia anacua* (anacua), possibly *Quercus fusiformis* (live oak), and others (Van Auken and Bush 1985). However, light and soil nitrogen requirements of most of these species are not known.

In drier areas of the American southwest, the proposed woody legume pioneer species or encroaching species would be different and could include *Senegalia berlandieri* (gaujillo), *Vachellia rigidula* (blackbrush acacia), *V. neovernicosa* (*A. neovernicosa*), *V. constricta* (*A. constricta*), and possibly others. Canopies would be more open with higher light levels below the canopy, but in addition, there would still be higher levels of soil nitrogen below the canopies (Jurena and Van Auken 1998; Powell 1998). Seedlings of the same mature canopy species were found below the canopy of *V. rigidula* and *S. berlandieri* in these communities, but at low density, and 17–24 other woody or succulent species were found below the canopies as well. Some species were only found below the woody legume canopies, but findings were not consistent. The canopies seemed to be important for some species either because of higher levels of nitrogen in the soil or lower light levels, but it was not obvious which species would be replacement species. The communities with these species of woody legumes are extensive in area covered in the American southwest but less than the area covered by *Prosopis*. These communities are more open with lower stature than the mature *V. farnesiana* communities that are in areas of higher rainfall and soil moisture.

The studies that we have just cited concerning encroachment and secondary succession are not temporal studies but spatial studies or studies where the species seem to fit into a successional sequence based on dendrochronological and other observations, but the communities that we studied were not really a temporally sequence as indicated above. These studies were used to develop a descriptive model of woody legume secondary succession in the southwestern North American

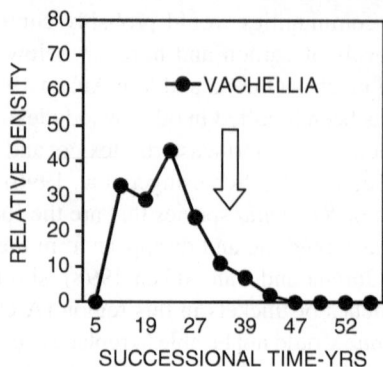

Fig. 4.5 The relative density of *Vachellia farnesiana* is presented in a space for time study (*left of arrow*) and a proposed or modeled chronosequence. The actual total woody plant density in these communities was 0–2,200 plants/ha (modified from Van Auken and Bush 1985). The actual successional time sequence for *V. farnesiana* measured in 2003 using some of the same communities sampled 20 years earlier is to the *right of the arrow*. The total density in these communities was approximately 1,300–2,200 plants/ha (modified from Bush et al. 2006) and then decreased to zero (*right of arrow*). These relative values show that the predicted trend from 1983 has continued

arid grasslands. They are considered space for time studies, and the successional stages in time are represented by community gradients in space (Begon et al. 2006). They are usually coupled with the use of various maps, including soil survey maps, aerial photographs, interviews with residents, and possibly dendrochronological or tree coring studies which are all tools that we used (Van Auken and Bush 1985).

After 20 years, we revisited some of these communities and demonstrated that the proposed encroachment and secondary succession based on space for time sampling and modeling have been confirmed as a true chronosequence of secondary succession that was based on temporal studies (Bush et al. 2006). In our first study (Van Auken and Bush 1985), we showed that total density of woody plants increased from zero to approximately 2,000 plants/ha and that total basal area of the woody plants increased from zero to approximately 20 m²/ha (data not shown). The two dominant species in the chronosequence, *V. farnesiana*, the proposed encroaching woody legume and early successional species, and *C. laevigata*, the late successional species, behaved as predicted in the earlier study. *Vachellia farnesiana* populations increased in density and relative density during the first 25 years of secondary succession then decreased in both density and relative density (Fig. 4.5). In addition, basal area and relative basal area of *V. farnesiana* increased in the early successional communities and then decreased in the communities that were resampled after 20 years (Fig. 4.6). These were communities that we previously estimated to be 19- to 32-years post-encroachment or agricultural abandonment, and when they were resampled, they were 39- to 52-years post-encroachment or abandonment. *Vachellia farnesiana* density, relative density, basal area, and relative basal area decreased to essentially zero in communities that were now 47- to 52-year post-agricultural abandonment. The density *of the* late successional species, *Celtis laevigata,* decreased

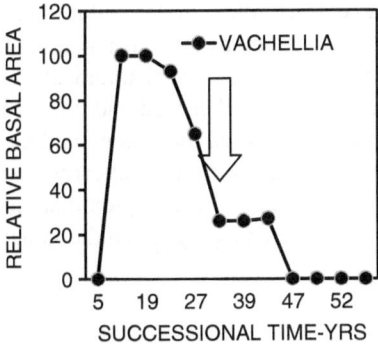

Fig. 4.6 The relative basal area of *Vachellia farnesiana* is presented in a space for time study (*left of arrow*) or modeled chronosequence. The actual total woody plant basal area in these communities was 0–24 m²/ha (modified from Van Auken and Bush 1985). The actual successional time sequence for *V. farnesiana* measured in 2003 using some of the same communities sampled 20 years earlier is to the *right of the arrow*. The total basal area in these communities was approximately 15–20 m²/ha (modified from Bush et al. 2006) and then decreased to zero (*right of arrow*). These relative values show that the predicted trend from 1983 has continued

somewhat, but basal area remained the same or increased a little. Both density and basal area of *C. laevigata* were similar to values reported for the oldest communities sampled 20 years previously and similar to the mature communities that we sampled earlier (data not shown).

In addition to the studies presented above to demonstrate secondary succession in these woody legume communities, we also used DCA community ordination. Community age and DCA axis 1 positions were highly correlated with either total density or total basal area ($r=-0.95$, $P \le 0.0001$ and $r=-0.97$, $P \le 0.0001$, respectively). Moreover, as the plant communities were resampled, new soil samples were collected and analyzed from the communities in the proposed chronosequence (Bush 2008). Total soil carbon and nitrogen continued to increase in these successional communities. Total soil carbon increased from about 30 g/kg to about 60 g/kg in the now 52-year-old successional communities. Total soil nitrogen increased from approximately 3 g/kg to approximately 5 g/kg. When all of the total soil carbon values including the mature community values (>150-year-old communities) were regressed on time, a quadratic function explained most of the variation. The $R^2 - 0.64$ with a $P \ge 0.0001$ further confirming the successional sequence. The same was true for total nitrogen. The $R^2 = 0.70$ with a $P \ge 0.0001$, again confirming the successional sequence.

Besides the increasing levels of total soil carbon and nitrogen, identified in our earlier studies, in 1983, we found differences between total soil carbon and nitrogen below the canopies of the 15-year encroachment or post-agricultural abandoned grassland or savanna with a few *V. farnesiana* trees and in the intervening intercanopy gaps or patches. Approximate mean value for total soil carbon was 12 g/kg in the intercanopy and 22 g/kg below the canopy. The approximate mean values for total soil nitrogen were 1.4 g/kg in the intercanopy gap and 2.5 g/kg below the canopy.

Both total soil carbon and total soil nitrogen values were significantly different between the canopy and the intercanopy patch (Student's t-test, $P \leq 0.05$). When the soils from 25-year-old encroached or post-agricultural abandoned communities were resampled, the differences in total soil carbon and nitrogen disappeared, and there were no significant differences between the canopy and the intercanopy patch (student's t-test, $P \geq 0.05$). Mean values for total soil carbon were 31 g/kg in the intercanopy and 34 g/kg below the canopy. The approximate mean values for total soil nitrogen were 3.0 g/kg in the intercanopy gap and the same below the canopy.

References

Archer S (1994) Woody plant encroachment into southwestern grasslands and savannas: rates, patterns and proximate causes. In: Vavra M, Laycock WA, Pieper RD (eds) Ecological implications of livestock herbivory in the west. Society for Range Management, Denver, pp 13–69

Archer S, Scifres C, Bassham CR, Maggio R (1988) Autogenic succession in a subtropical savanna: conversion of grassland to thorn woodland. Ecol Monogr 52:111–127

Begon M, Townsend CR, Harper JL (2006) Ecology: from individuals to ecosystems, 4th edn. Blackwell, Malden, MA

Browning DM, Archer S (2011) Protection from livestock fails to deter shrub proliferation in a desert landscape with a history of heavy grazing. Ecol Appl 21:1629–1642

Bush JK (2008) Soil nitrogen and carbon after twenty years of riparian forest development. Soil Sci Soc Am J 72:815–822

Bush JK, Van Auken OW (1986a) Light requirements of *Acacia-smallii* and *Celtis-laevigata* in relation to secondary succession on floodplains of South Texas. Am Midl Nat 115:118–122

Bush JK, Van Auken OW (1986b) Changes in nitrogen, carbon, and other surface soil properties during secondary succession. Soil Sci Soc Am J 50:1597–1601

Bush J, Van Auken OW (1987a) Some demographic and allometric characteristics of *Acacia smallii* (Mimosaceae) in successional communities. Madrono 34:250–259

Bush JK, Van Auken OW (1987b) Light requirements for growth of *Prosopis-glandulosa* seedlings. Southwestern Nat 32:469–473

Bush JK, Van Auken OW (1989) Soil resource levels and competition between a woody and herbaceous species. Bull Torrey Bot Club 116:22–30

Bush JK, Van Auken OW (1991) Importance of time of germination and soil depth on growth of *Prosopis-glandulosa* (leguminosae) seedlings in the presence of a C-4 grass. Am J Bot 78:1732–1739

Bush JK, Van Auken OW (1995) Woody plant-growth related to planting time and clipping of a C-4 grass. Ecology 76:1603–1609

Bush JK, Richter FA, Van Auken OW (2006) Two decades of vegetation change on terraces of a south Texas river. J Torrey Bot Soc 133:280–288

Heitschmidt RK, Struth JW (1991) Grazing management: an ecological perspective. Timberline, Portland

Jurena PN, Van Auken OW (1998) Woody plant recruitment under canopies of two acacias in a southwestern Texas shrubland. Southwestern Nat 43:195–203

Lohstroh RJ, Van Auken OW (1987) Comparison of canopy position and other factors on seedling growth in *Acacia Smallii*. Texas J Sci 39:233–239

Powell AM (1998) Trees and shrubs of the Trans-Pecos and adjacent areas. University of Texas Press, Austin

Schlesinger WH, Raikes JA, Hartley AE, Cross AF (1996) On the spatial pattern of soil nutrients in desert ecosystems. Ecology 77:364–374

Scifres CJ (1980) Brush management: principles and practices for Texas and the Southwest, 1st edn. Texas A&M University Press, College Station, TX

Simpson BB (1977) Mesquite: its biology in two desert ecosystems. Dowden, Hutchingson & Ross, Inc, Stroudsburg, PA

Smith HN, Rechenthin CA (1964) Grassland restoration: the Texas brush problem. United States Department of Agriculture, Soil Conservation Service

Taylor CA Jr, Twidwell D, Garza NE, Rosser C, Hoffman JK, Brooks TD (2012) Long-term effects of fire, livestock herbivory removal and weather variability in Texas semiarid savanna. Rangeland Ecol Manage 65:21–30

Tilman D (1985) The resource-ratio hypothesis of plant succession. Am Nat 125:827–852

Van Auken OW (2000) Shrub invasions of North American semiarid grasslands. Annu Rev Ecol Syst 31:197–215

Van Auken OW (2009) Causes and consequences of woody plant encroachment into western North American Grasslands. J Environ Manage 90:2931–2942

Van Auken OW, Bush JK (1985) Secondary succession on terraces of the San Antonio River. Bull Torrey Bot Club 112:158–166

Van Auken OW, Bush JK (1989) *Prosopis glandulosa* growth—influence of nutrients and simulated grazing of *Bouteloua curtipendula*. Ecology 70:512–516

Van Auken OW, Bush JK (1997) Growth of *Prosopis glandulosa* in response to changes in aboveground and belowground interference. Ecology 78:1222–1229

Van Auken OW, Gese EM, Connors K (1985) Fertilization response of early and late successional species: *Acacia smallii* and *Celtis laevigata*. Bot Gazette 146:564–569

Chapter 5
Factors that Determine Growth Rates

The chronological appearance and function of the woody legumes in many of the C_4 grassland communities is still not completely understood or accepted by all ecologists or rangeland scientists. *Vachellia*, particularly *V. farnesiana* and *V. rigidula, as well as Senegalia berlandieri* and *Prosopis glandulosa* appear to be early successional species that establishes on low nutrient or low nitrogen soils (Van Auken et al. 1985) or in grassland communities that have had high levels of disturbances via extensive continuous domestic grazing and no or few fires or probably both (Archer et al. 1988; Van Auken 2009). Two factors or conditions that may be limiting and that are very important for the establishment and growth of these species or community are surface light levels and soil nitrogen levels. These are factors suggested by Tilman as potentially responsible for the sequence of species in the resource ratio theory of community development in succession (Tilman 1985).

The growth of *V. farnesiana* was not promoted by nitrogen addition to the soil, but its growth was promoted by the addition of other nutrients (Fig. 5.1) (Van Auken et al. 1985; Van Auken 1994). If we are truly demonstrating secondary succession, then there should be a species that takes the place of *V. farnesiana* during secondary succession. In this south central Texas successional sequence, we have shown that the species is *Celtis laevigata* (hackberry or sugar hackberry). If the resource ratio hypothesis is true and represents what is occurring in this successional sequence, *C. laevigata* should respond differently to added soil resources. *Celtis laevigata* was not stimulated by additional added soil nutrients, but its growth increased 5.6 times with added soil nitrogen (Fig. 5.1).

A second environmental factor that seems to be important in secondary succession is light level. During succession light levels near the surface of the soil will decrease with the development of a community and a canopy (Fig. 3.1). This is part of the resource ratio theory of succession. Early successional species would be high light requiring species or sun species, while late successional species would be low light requiring species or shade species. Light requirements of *V. farnesiana,* a proposed early successional sun species, and *C. laevigata,* a proposed late successional shade species, were tested and compared.

O.W. Van Auken and J.K. Bush, *Invasion of Woody Legumes*, SpringerBriefs in Ecology 4, 25
DOI 10.1007/978-1-4614-7199-8_5, © Springer Science+Business Media New York 2013

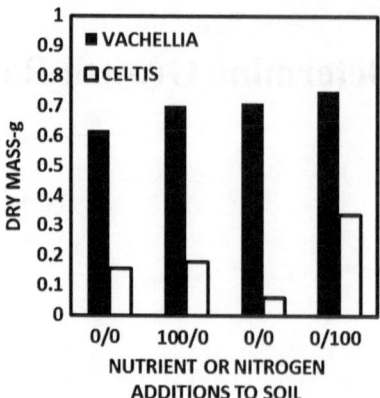

Fig. 5.1 Dry mass production or growth of *Vachellia farnesiana* and *Celtis laevigata* over 56 days in low nutrient Frio clay-loam soil with added nutrients or added nitrogen (modified from Van Auken et al. 1985). (*Vachellia farnesiana* one-way ANOVA for nutrient addition, $F = 2.78$, $P < 0.009$, for nitrogen addition, NSD; *Celtis laevigata* one-way ANOVA for nutrient addition, NSD, one-way ANOVA for nitrogen addition, $F = 3.85$, $P < 0.0001$)

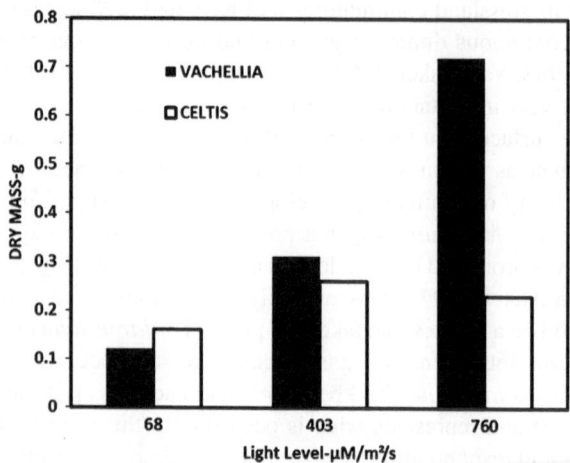

Fig. 5.2 Mean dry mass of *Vachellia farnesiana* and *Celtis laevigata* at three light levels from 68 ± 20 µM/m²/s to 760 ± 96 µM/m²/s in Frio clay-loam soil with added soil resources. Dry mass of *V. farnesiana* was linearly related to light levels ($r = 0.99$, $P < 0.001$), while *Celtis laevigata* was not ($r = 0.79$, $P > 0.05$) (modified from Bush and Van Auken 1986)

Vachellia farnesiana growth is promoted by high levels of visible light suggesting that it is a sun plant or sun species and its growth is suppressed in low light and in its own shade below its own canopy or by the shade of other canopy species (Bush and Van Auken 1986; Lohstroh and Van Auken 1987). *Vachellia farnesiana* dry mass increased linearly with increased light levels ($r = 0.99$, $p \geq 0.001$) (Fig. 5.2).

A species that takes its place during secondary succession is *C. laevigata*. It appears to be a shade or shade-tolerant species, and its dry mass did not increase linearly with increased light levels ($r=0.79$, $p>0.05$), but reached a plateau at low light levels (Fig. 5.2). *Vachellia farnesiana* seedlings are found in open disturbed areas and open grassland associated with supposed parent trees (partial shade) early in succession (Fig. 4.3), but they are not present when the canopy closes with community development probably because of low light levels below the canopy. The reverse is true for *C. laevigata* (Fig. 4.3), and its growth is promoted and not suppressed in low light (Bush and Van Auken 1986).

Transplant studies have been done with both *V. farnesiana* and *C. laevigata* (Lohstroh and Van Auken 1987; Van Auken and Lohstroh 1990). Both species were germinated from seed and placed in a Frio clay-loam low nutrient soil (Taylor et al. 1966). Plants of both species were grown in a greenhouse for seven weeks before being transplanted into the field site. Plants were grown in 15 cm deep plastic pots containing 1,400 g of dried and sieved soil. The experiment was a factorial experiment, and plants were placed under the canopy of mature *V. farnesiana* trees or in gaps between the trees. Nutrients were added or not added, roots of other species were present or removed via trenching to a 20 cm depth, and insect herbivores were present or removed via the addition of insecticide. Mean light levels below the canopy were 515 ± 153 μM/m²/s, and in the gaps, they were $2,173 \pm 174$ μM/m²/s on a clear midday, measured with a Li-cor® LI-188 integrating quantum sensor. The experiment was harvested on the first of November which was 12 weeks after transplanting into the field. The only significant differences between treatments were for canopy position. *Vachellia farnesiana* aboveground dry mass in the open or gaps was approximately twice as high as the dry mass below the *V. farnesiana* canopy with none of the other treatments showing significant differences. The reverse was true for *C. laevigata, and* aboveground dry mass below the *V. farnesiana* canopy was approximately four times higher than dry mass in the open or gaps with none of the other treatments showing significant differences.

The appearance and function of *Prosopis glandulosa* (Archer et al. 1988) in former arid and semiarid C_4 temperate, subtropical, and tropical grassland seems very similar to what is reported for *V. farnesiana*. *Prosopis* species appear to be early successional species that encroach and establish on low nutrient or low nitrogen soils or in disturbances in grasslands or savannas with constant, heavy domestic grazing (Archer et al. 1988; Archer 1994). *Prosopis* will grow in low nutrient soils, but its growth is promoted by nutrient addition to the soil, including N, P, K, and S (Fig. 5.3) (Van Auken and Bush 1989). The greatest stimulation was with the addition of all four nutrients.

The importance of high rates of biological nitrogen fixation in early successional or disturbed communities and the role of legumes, like *Prosopis*, in these communities have been known for a considerable time (West and Klemmedson 1978). Soil resource levels, especially nitrogen, below a *Prosopis juliflora* canopy in the Sonoran desert were higher than values in gaps or intervening areas between canopies (West and Klemmedson 1978). Total nitrogen levels in the intervening gaps between *Prosopis* plants were 0.02–0.04 %, while at the surface below the *Prosopis*

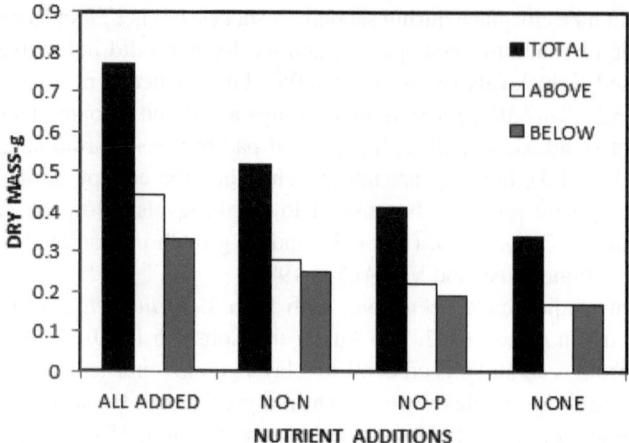

Fig. 5.3 Mean total, aboveground, and belowground dry mass of *Prosopis glandulosa* with complete nutrient addition (*left*), minus nitrogen, minus phosphorus, and a control with no added nutrients(*right*) (Van Auken and Bush 1989). The ANOVA for total, aboveground, and belowground dry mass was significant ($P<0.0001$ for each variable), with the complete media (all added) being significantly different from the control (no added nutrients), but not different from the No-N or No-P added nutrients (Scheffé Multiple Range Test, $P>0.05$)

canopy, total nitrogen values were approximately 0.06 %. Total soil nitrogen levels were higher below the *Prosopis* canopy compared to the surrounding gaps or open area soils to a depth of 20–25 cm.

Total soil carbon and nitrogen levels approximately doubled over a period of 20 years in *Vachellia* communities a little farther north and east (Bush 2008). Similar trends were found in other woody legume communities with soil nitrogen below the canopy of *S. berlandieri* and *V. rigidula* being 5 and 13 times higher compared to gaps or open areas between the plants. There are similar findings for soil nitrogen levels in other parts of the arid regions of southwestern North America including areas in both the Chihuahuan and Sonoran deserts (Virginia and Jarrell 1983; Virginia 1986; Schlesinger et al. 1996).

Lower light levels below the *Prosopis* canopy have also been demonstrated (Bush and Van Auken 1987). Light levels below the *Prosopis glandulosa* canopy above the herbaceous layer were 20–45 % of full light levels in associated gaps. Below the herbaceous layer, light levels were further reduced to 10–25 % of the full light levels (Fig. 5.4).

There were significant differences in light levels above the surface vegetation and below the surface vegetation in the open compared to under the canopy (one-way ANOVA for both positions, $P<0.001$). *Prosopis glandulosa* growth is promoted by high levels of visible light suggesting that it is a sun not a shade species and its growth is suppressed in its own shade below its own canopy or by the shade of other canopy species (Bush and Van Auken 1987). Its dry mass increased linearly with increased light levels over the range of light levels examined ($r^2=0.99$, $P\geq0.001$) (Fig. 5.5).

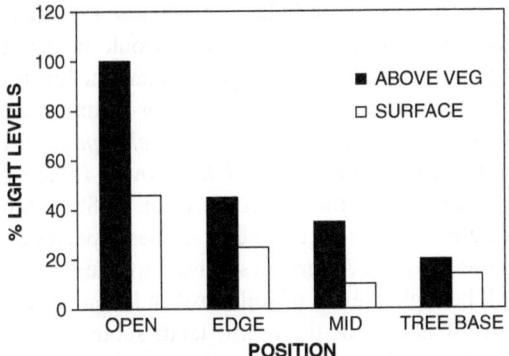

Fig. 5.4 Percent light levels both above and below surface herbaceous vegetation in the open grassland or gaps (*left*), at the *Prosopis glandulosa* canopy edge, at the canopy midpoint, and at the base of the *P. glandulosa* tree (*right*). The mean light level in the open above the herbaceous vegetation was $1{,}858 \pm 139$ μmol m^{-2} s^{-1} (Bush and Van Auken 1987)

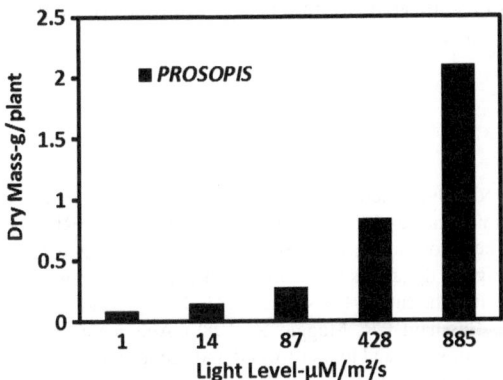

Fig. 5.5 Mean dry mass of *Prosopis glandulosa* grown at five light levels from <1.0 μM/m^2/s to 885 μM/m^2/s in a Patrick series clay-loam soil with added nutrients. Dry mass of *Prosopis glandulosa* was linearly related to light levels ($r=0.99$, $P<0.001$) (modified from Bush and Van Auken 1987)

Prosopis glandulosa seedlings (small plants or juveniles) are found in open disturbed areas or open grassland early in succession, but they do not seem to be present below the canopy as the community develops probably because of low light levels below the canopy (Fig. 5.5) (Bush and Van Auken 1987; Archer et al. 1988).

In time the *P. glandulosa plants* that are the focus of these patches or clusters in woodlands or savannas would die and disappear as in succession (Archer et al. 1988; Archer 1994). *Prosopis glandulosa* would not be able to replace itself in woodland patches probably because of lower light levels below its canopy (Bush and Van Auken 1987; Archer 1995). Replacement species would be any number of trees or shrubs capable of growth in the higher nitrogen soil and the low light environment of the canopy understory. Species identified in older or mature *Prosopis* communities included *Zanthoxylum fagara* (lime prickly ash), *Celtis pallida* (desert hackberry), *Diospyros texana* (Texas persimmon), *Condalia obovata* (blue wood), *Berberis trifoliolata* (agarito), and a few others (Archer et al. 1988). However, the growth requirements especially the light requirements of these species are not well defined at this time. In drier areas of the American southwest, canopies would be more open, but there would still be higher levels of soil nitrogen below the canopies.

Apparently many if not all of these arid land, subtropical, or tropical woody legumes can or have been able to encroach and establish in area grasslands that have been disturbed by heavy and continuous domestic grazing and a reduced fire frequency. Once the woody plants are established, the communities would go through secondary succession responding to soil resource levels and surface light levels. The temporal sequence of species change in the communities would be from grassland to savanna to woodland with the late stage or stages undefined. The temporal sequence will be governed by the amount of annual rainfall and probably other factors. The time it takes to reach a late stage in succession will vary by approximately 150 years in fairly wet areas (75 cm/year) to an unknown number of years in drier regions.

References

Archer S (1994) Woody plant encroachment into southwestern grasslands and savannas: rates, patterns and proximate causes. In: Vavra M, Laycock WA, Pieper RD (eds) Ecological implications of livestock herbivory in the West. Society for Range Management, Denver, pp 13–69

Archer S (1995) Tree-grass dynamics in a *Prosopis*-thronscrub savanna parkland: Reconstructing the past and predicting the future. Ecoscience 2:83–99

Archer S, Scifres C, Bassham CR, Maggio R (1988) Autogenic succession in a subtropical savanna: conversion of grassland to thorn woodland. Ecol Monogr 52:111–127

Bush JK (2008) Soil nitrogen and carbon after twenty years of riparian forest development. Soil Sci Soc Am J 72:815–822

Bush JK, Van Auken OW (1986) Light requirements of Acacia-smallii and Celtis-laevigata in relation to secondary succession on floodplains of south Texas. Am Midl Nat 115:118–122

Bush JK, Van Auken OW (1987) Light requirements for growth of Prosopis-glandulosa seedlings. Southwestern Nat 32:469–473

Lohstroh RJ, Van Auken OW (1987) Comparison of canopy position and other factors on seedling growth in *Acacia Smallii*. Texas J Sci 39:233–239

Schlesinger WH, Raikes JA, Hartley AE, Cross AF (1996) On the spatial pattern of soil nutrients in desert ecosystems. Ecology 77:364–374

Taylor FB, Hailey RB, Richmond DL (1966) Soil survey of Bexar County, Texas. USDA, Soil Conservation Service, Washington

Tilman D (1985) The resourse-ratio hypothesis of plant succession. Am Nat 125:827–852

Van Auken OW (1994) Changes in competition between a C_4 grass and a woody legume with differential herbivory. Southwestern Nat 39:114–121

Van Auken OW (2009) Causes and consequences of woody plant encroachment into western North American Grasslands. J Environ Manage 90:2931–2942

Van Auken OW, Bush JK (1989) Prosopis-glandulosa growth—influence of nutrients and simulated grazing of Bouteloua-curtipendula. Ecology 70:512–516

Van Auken OW, Lohstroh RJ (1990) Importance of canopy position for the growth of *Celtis laevigata* seedlings. Texas J Sci 42:83–89

Van Auken OW, Gese EM, Connors K (1985) Fertilization response of early and late successional species: *Acacia smallii* and *Celtis laevigata*. Bot Gazette 146:564–569

Virginia RA (1986) Soil development under legume canopies. For Ecol Manage 16:69–79

Virginia RA, Jarrell WM (1983) Soil properties in a mesquite-dominated Sonoran desert ecosystem. Soil Sci Soc Am J 47:138–144

West NE, Klemmedson JO (1978) Structural distribution of nitrogen in desert ecosystems. In: West NE, Skujins JJ (eds) Nitrogen in desert ecosystems. Dowden, Hutchinson and Ross Inc, Stroudsberg, PA

Chapter 6
Competition

The interactions between individuals within species populations or competition have been considered one of the major factors controlling or determining community characteristics (Harper 1977; Grime 1979; Grace and Tilman 1990). The interactions occur because of shared requirements for the same resource and can lead to negative, positive, or no effect on one or both individuals, species, or populations (Begon et al. 2006). However, the topic has been continually debated (Grace 1995; Twolan-Strutt and Keddy 1996; Van Auken 2009). We have done a number of greenhouse-controlled environment competition studies and field studies between *V. farnesiana*, *P. glandulosa*, and various grasses from former C_4 grasslands to help to understand reasons for the changes that have occurred with encroachment of woody plants and long-term, constant, and high levels of herbivory in these grasslands (Cohn et al. 1989; Van Auken 1994).

In one experiment, *V. farnesiana* was grown with a C_4 grass (*Cynodon dactylon* Bermuda grass) which is an introduced Eurasian grass that is now ubiquitous in Texas and many C_4 grasslands (Gould 1969; Correll and Johnston 1979). The experiment was carried out in pots, in a low nutrient soil (clay-loam Mollisol of the Frio series) (Taylor et al. 1966). Little growth per plant occurred for either species in mixtures of the two species or monoculture (by themselves) in low nutrient soil, but there was more grass (Fig. 6.1). In addition, there were no significant differences in dry mass between proportional treatments between mixture and monoculture for either species (Cohn et al. 1989). Relative yields were higher for *C. dactylon* in mixture (30–50 %) and were lower for *V. farnesiana* in mixtures (10–20 %), but neither was significantly different. When nutrients were added, growth of *C. dactylon* doubled in monoculture and tripled in 2:4 mixtures (*Cynodon/Vachellia*), and the growth of *V. farnesiana* decreased in competition with the grass, but not in monoculture where it increased slightly (Fig. 6.1) (Cohn et al. 1989). *Cynodon* dry mass was higher than the *V. farnesiana* dry mass in all cases. Both species appear to be early successional colonizers and probably germinate and start growth in gaps caused by disturbances. *Cynodon* was the better competitor in the soils with added nutrient, but the species were approximately equal in low nutrient soil in spite of greater grass dry mass. In addition, there were no mortalities of either species.

O.W. Van Auken and J.K. Bush, *Invasion of Woody Legumes*, SpringerBriefs in Ecology 4, DOI 10.1007/978-1-4614-7199-8_6, © Springer Science+Business Media New York 2013

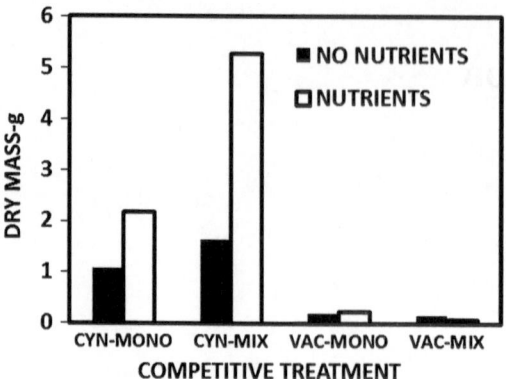

Fig. 6.1 Mean dry mass per plant of *Cynodon dactylon* (CYN=Bermuda grass) and *Vachellia farnesiana* (VAC) grown without nutrients and with nutrients. Plants were grown in monoculture by themselves with six plants per pot (MONO) or in mixture (MIX-ratio of 4:2) *Cynodon* to *Vachellia* or the reverse (2:4) *Vachellia* to grass (modified after Cohn et al. 1989). A three-way ANOVA was used to analyze dry mass per plant with nutrient level (2), proportion (5), and species (2) as main effects and all first and second-order interactions. Nutrient level, proportion, and species were all significant in the ANOVA ($P < 0.05$)

When simulated herbivory (clipping one or the other species) was added to a similar competition experiment with the same species both with and without nutrients, the results were complex but reveling (Van Auken 1994). Main experimental treatments were significant in the three-way factorial experiment for the grass (*Cynodon dactylon*) with significant interactions. Thus, for the grass, dry mass harvested was dependent on the nutrient level (added or not added), herbivory treatment (grass clipped or not clipped), and in some cases the condition of the neighbor (present, but clipped or not). When nutrients were added to the soil, dry mass of *C. dactylon* grown alone but unclipped increased almost three times from 5 g/pot aboveground dry mass to about 15 g/pot (not shown). Simulated herbivory or clipping in low nutrient soil had a small, negative effect on grass dry mass production, but with nutrients added and clipping, there was an increase in dry mass. There was little or no effect on grass dry mass when one *V. farnesiana* plant was grown with the grass with or without nutrient addition or simulated herbivory (not shown).

Effects on *V. farnesiana* were quite different. Some of the main experimental treatments were significant in the three-way factorial experiment for the woody legume with some significant interactions. When nutrients were added to the soil, dry mass production of *V. farnesiana* grown alone and unclipped was slightly greater compared to dry mass when no nutrients were added (not significant, Fig. 6.2). When *V. farnesiana* was clipped, the clipped plants were 3–5 % of the non-clipped plants with added nutrients having a small effect (Fig. 6.2). When *V. farnesiana* was grown with the grass, woody legume dry mass was reduced 95–98 % if the grass was not clipped. If the grass was clipped, but nutrients were added, *V. farnesiana* dry mass was suppressed 60 % compared to growth alone and not clipping the woody plant (left panel). In the same treatment without added nutrients (right panel), *V. farnesiana* was suppressed 96 % (Fig. 6.2) (Van Auken 1994).

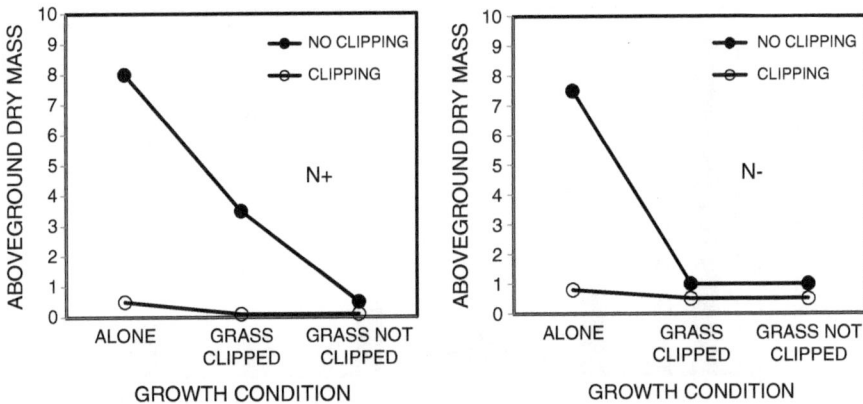

Fig. 6.2 Aboveground dry mass of *Vachellia* is presented. Dry mass was measured with or without competition and herbivory between *Vachellia farnesiana* a C₃ woody plant and *Cynodon dactylon* a C₄ grass. In the right panel, no nutrients were added, and in the *left panel*, nutrients were added. The target species (*Vachellia*) was not clipped or clipped. In addition, the target species (*Vachellia*) was grown in three competition treatments, ALONE or with the grass competitor that was either clipped (GRASS CLIPPED) or not clipped (GRASS NOT CLIPPED), note the *x*-axis (modified from Van Auken 1994)

Similar greenhouse competition studies were done between *P. glandulosa* and C₄ grasses from these former C₄ arid and semiarid grasslands (Van Auken and Bush 1987, 1988, 1989, 1990, 1997; Bush and Van Auken 1991, 1995). When one *P. glandulosa* plant was grown separately with *Schizachyrium scoparium* little blue-stem or *Bouteloua curtipendula* side oats grama in greenhouse additive experiments, in low nutrient soil, dry mass or growth of the woody plant was dependent on the grass density of both grass species. The presence of one grass plant reduced the aboveground dry mass of the *P. glandulosa* plants by 46 or 84 %, depending on the species it was grown with, and belowground dry mass by 67 or 69 % (Fig. 6.3 for *S. scoparium*) (Van Auken and Bush 1988; Bush and Van Auken 1989). At the highest grass density tested, 97 *S. scoparium* plants per pot, *P. glandulosa* total dry mass was approximately 4 % of the control (no grass present, Fig. 6.3). However, there were no *Prosopis* mortalities. *Prosopis glandulosa* was suppressed by 96 % at the high grass density but not killed.

An interspecific (de Wit) greenhouse competition experiment (Harper 1977) with and without addition of nutrients was completed with *P. glandulosa* and *B. curtipendula* at a density of six plants per pot. Proportions of the two species were 6:0, 4:2, 3:3, 2:4, and 0:6, respectively. The results were the same at both nutrient levels; the grass was always the better competitor (Fig. 6.4) (Van Auken and Bush 1988, 1989; Bush and Van Auken 1989). Relative yields per plant for all proportions of *P. glandulosa* and *B. curtipendula* were significantly different from the relative yield for the plant without a competitor. Relative yield totals for *B. curtipendula* without the addition of nutrients in mixture were 80–92 % depending on the proportion of plants and 95–98 % with nutrients added but depending on the proportion of plants present. The percent yield for *P. glandulosa* was 8–20 % without added nutrients, and it

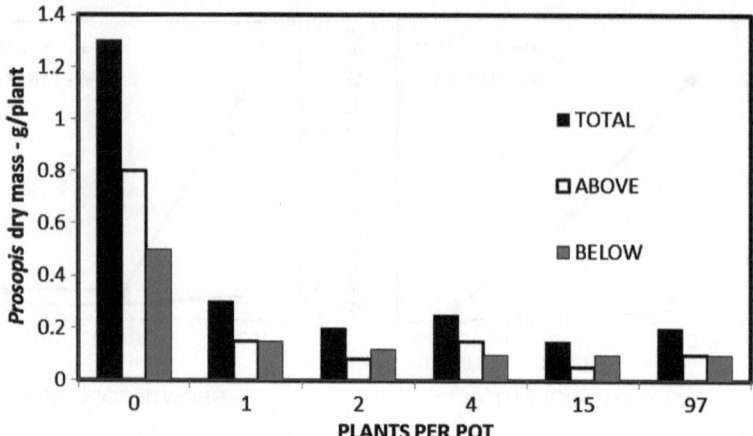

Fig. 6.3 Mean total, aboveground, and belowground dry mass of one *Prosopis glandulosa* plant grown with *Schizachyrium scoparium* (little bluestem) in an additive experiment (modified from Bush and Van Auken 1989). A two-way ANOVA with treatment (grass density) and replication (block) followed by the least significant difference test was completed. There was no significant difference in replication ($P > 0.05$) and grass densities of 1–97 were all significantly different from the zero grass density but not each other ($P < 0.05$, LSD)

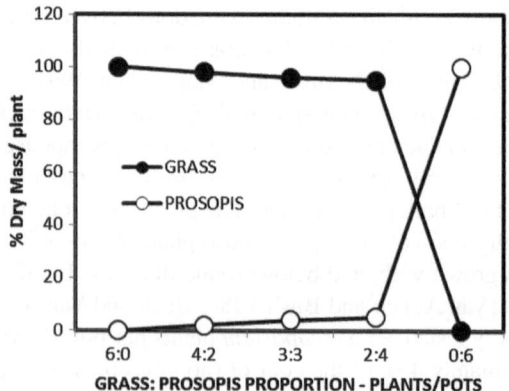

Fig. 6.4 Mean relative yields of *Prosopis glandulosa* and *Bouteloua curtipendula* (grass) grown at a density of six plants per pot (in native Patrick soil, no added nutrients) but at various proportions including 6:0, 4:2, 3:3, 2:4, and 0:6 (modified from Bush and Van Auken 1989). The one-way ANOVA was significant and followed by the Scheffé multiple comparison test $P < 0.05$

was reduced to 2–5 % with added nutrients also dependent on the proportion of plants present, but there were no mortalities (Bush and Van Auken 1989).

When simulated herbivory of *B. curtipendula* was added to a competition experiment with and without nutrients, the results were very telling (Van Auken and Bush 1989). Aboveground, belowground, and total dry grass mass was highest with complete nutrients (N, P, K, and S) and reduced when any one nutrient was not added,

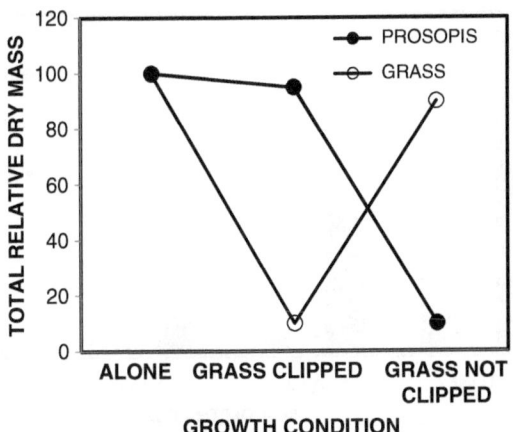

Fig. 6.5 Total relative dry mass of one *Prosopis glandulosa* plant grown alone, or with the C₄ grass *Bouteloua curtipendula* that was clipped, or with *B. curtipendula* that was not clipped (modified from Van Auken and Bush 1989). One-way ANOVA for total, aboveground, and belowground dry mass for the *Prosopis* or grass dry mass ($P < 0.0001$) followed by the Scheffé multiple comparison test showed that there was not a significant difference ($P > 0.05$) between all parameters when *Prosopis* was grown alone or when the grass was clipped. However, both the Prosopis relative dry mass when grown alone and when the grass was clipped were different from the grass-no clip treatment ($P < 0.05$)

which was the same for *P. glandulosa* (Fig. 5.3). When the grass plants were clipped, in the treatment with added nutrients, the grass total dry mass was reduced from 21.5 g/pot to 3.0 g/pot or 86 %. The grass total dry mass was 7.2 times greater without clipping. In addition, the clipping reduced the grass belowground dry mass from 9.5 g/pot to 0.5 g/pot or by 94.8 %. There was no significant difference in the grass dry mass when one *P. glandulosa* was planted in the same pot with the grass (data not shown). When *P. glandulosa* was planted with the grass and the grass was not clipped, *P. glandulosa* total dry mass was reduced from 0.80 g/plant (100 %) when grown alone with added nutrients to 0.12 g/plant when grown with the grass (Fig. 6.5). This was a 90 % reduction in total *P. glandulosa* dry mass and an 89 % reduction in below ground dry mass. However, when *P. glandulosa* was grown with the grass that was clipped with full nutrients added to the soil, there was no significant difference in total, aboveground, or belowground *P. glandulosa* dry mass. *Prosopis glandulosa* dry mass or growth when the grass was clipped to simulate grazing was the same as when the grass was not present (Fig. 6.5).

Early herbivory of *P. glandulosa* in a grassland or simulated herbivory has not been examined directly but should be. In North American grasslands, and probably worldwide grasslands, woody seedling defoliation and total aboveground plant consumption by rodents, lagomorphs and insects, or other herbivores is probably an important source of mortality (Van Auken 2009). We anticipate that results with *P. glandulosa* would be similar to simulated herbivory of *Vachellia farnesiana* as reported above. In fact, we suspect that herbivory of woody plant seedlings in intact

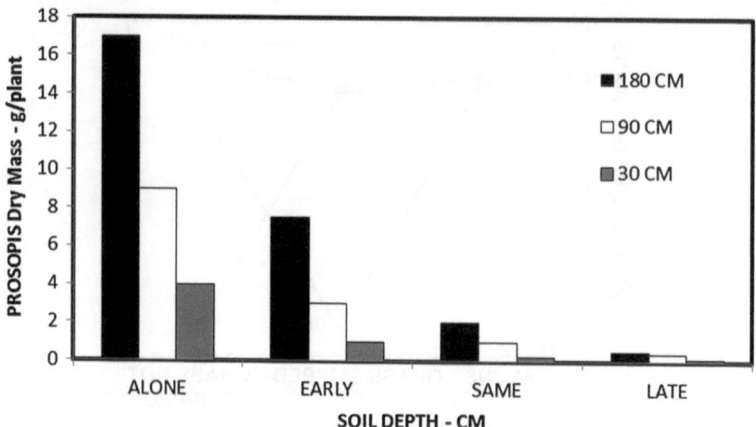

Fig. 6.6 Figure represents total dry mass (aboveground plus belowground) of *P. glandulosa* grown alone or with a C₄ grass Bouteloua *curtipendula*. Both *P. glandulosa* and the grass were planted in 180 cm, 90 cm, or 30 cm deep, 10 cm diameter PVC pipes (*pots*). *Prosopis glandulosa* was planted alone or with the grass, but planted 60 days before the grass was planted (early), at the same time as the grass, or 60 days after the grass was planted (late) (modified from Bush and Van Auken 1991)

native grasslands would be high as would be mortality and populations or potential populations of woody plants would be reduced or maintained relatively low. The previous comments concern aboveground woody plant herbivory, but there is no information available concerning belowground herbivory and its potential effects on woody plant growth, survival, or mortality.

Soil depth is another important environmental factor for dry mass production and growth of some arid land woody legumes like *P. glandulosa* (Bush and Van Auken 1991) which does have very deep roots (Van Auken 2000). When *P. glandulosa* was grown alone in 180 cm deep pots (10 cm diameter PVC pipes), *P. glandulosa* dry mass was 4.25 times greater compared to growth in 30 cm deep pots (Fig. 6.6). But, soil depth was not the only critical factor examined. The results of soil depth and relative time of planting were examined simultaneously in the experiment.

The woody legume was planted early (before the grass as in a gap), at the same time as the grass or after the grass. Results of the two-way ANOVA for aboveground, belowground, and total dry mass for *P. glandulosa* and both experimental factors (soil depth and planting combination) and their interaction were significant ($P < 0.0003$ for all). When grown in competition with *Bouteloua curtipendula*, pot depth (30, 90, or 180 cm) and the time of planting of *P. glandulosa* compared to the grass were both important (Fig. 6.6). When *P. glandulosa* was planted with the C₄ grass, but 60 days before the grass was planted (*P. glandulosa* planted early), *P. glandulosa* dry mass was reduced by approximately 50 % (compared to alone), but it was reduced less in the 180 cm deep pots. If *P. glandulosa* and the grass were planted at the same time, growth of the woody plant was further reduced compared to growth alone, but growth was still greater in the deeper pots. When *P. glandulosa* and the grass were planted together, but the woody plant was planted after the grass, the dry mass or growth of the woody plant was reduced to 3 % of dry mass when

Fig. 6.7 Mean *Bouteloua curtipendula* belowground dry mass/100 cm² extracted from 40 cm deep root cores from an experimental *B. curtipendula* grassland. Lengths of the 10 cm diameter PVC root excluders were 2, 20, and 40 cm. One-way ANOVA for belowground dry mass was significant ($P<0.05$)

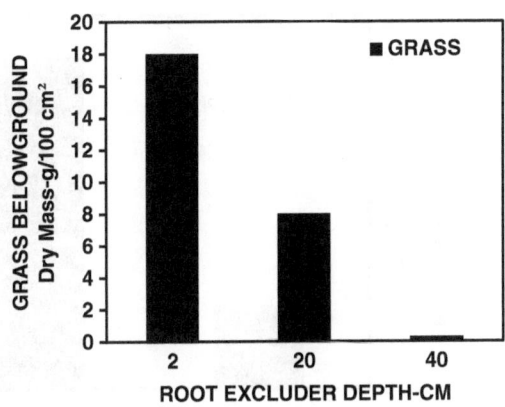

grown alone in the deep pots when it was planted early and there was no difference in the dry mass in the pots with different depths (Fig. 6.6). Dry mass of the grass, *Bouteloua curtipendula,* was dependent on the pot depth but not the time of planting or the presence of the woody plant. Results of the two-way ANOVA for aboveground, belowground, and total dry mass for *Bouteloua curtipendula* and the same experimental factors (soil depth and planting combination) and their interaction showed that soil depth was significant, but planting combination and their interaction was not ($P<0.0001$, 0.3869, and 0.0830, respectively, with values for belowground and total dry mass being very similar).

In the field, root excluders were used to examine potential belowground grass interference with woody legume growth, specifically *P. glandulosa* (Van Auken and Bush 1997). Root excluders were 10 cm diameter PVC pipes cut to lengths of 2, 20, and 40 cm and pounded into the ground until only the surface edge was exposed. The 2 cm excluder was essentially a surface control. When 20 cm long root excluders were used in *B. curtipendula* experimental grass monocultures, grass root dry mass in the excluders was reduced 58 % to 8 g/100 cm² compared to the 2 cm excluders (Fig. 6.7). When 40 cm root excluders were used, grass dry mass was reduced to almost zero or a trace. *Prosopis glandulosa* total dry mass in the 2 cm deep root excluders for one plant grown for one growing season was 0.2 g/plant and increased four times to 0.8 g/plant in the 20 cm root excluders with only a slight additional increase in the 40 cm deep excluders (Fig. 6.8).

The importance of understanding the potential effects of belowground and aboveground interference of a C_4 grass (*B. curtipendula*) on an encroaching woody legume like *P. glandulosa* was examined in the same experimental grassland as above (Van Auken and Bush 1997). The experimental treatments were *B. curtipendula* present as grassland or absent as in a gap, low or high light (with or without shading), and high grass roots (2 cm root excluder) and/or low grass roots (40 cm deep root excluders). The growth or total dry mass of *P. glandulosa* was measured at the end of one growing season after 7 months of growth. Dry mass of *P. glandulosa* was lowest at approximately 0.1 g/plant in the shade (low light), with high grass roots (the 2 cm root excluders). Highest *P. glandulosa* total dry mass was in the gaps with high light and low grass roots (Fig. 6.9). A significant, inverse,

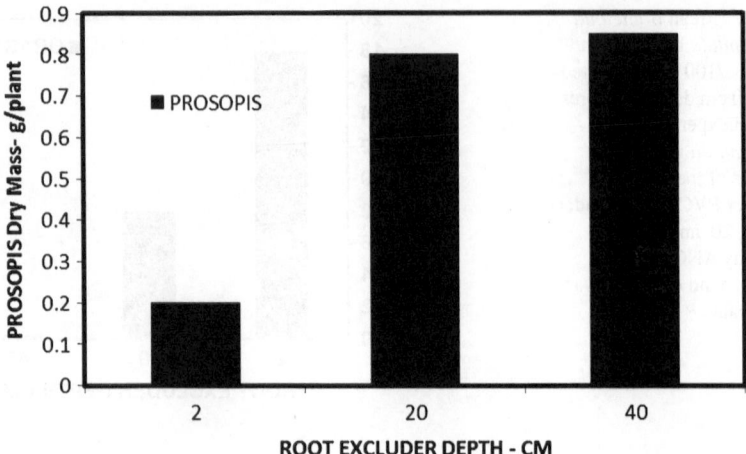

Fig. 6.8 Mean *Prosopis glandulosa* total dry mass per plant from three root excluders of different lengths. Root mass was extracted from 40 cm deep soil cores from an experimental *B. curtipendula* grassland. Lengths of the 10 cm diameter PVC root excluders were 2, 20, and 40 cm. One-way ANOVA for belowground dry mass was significant ($P<0.05$)

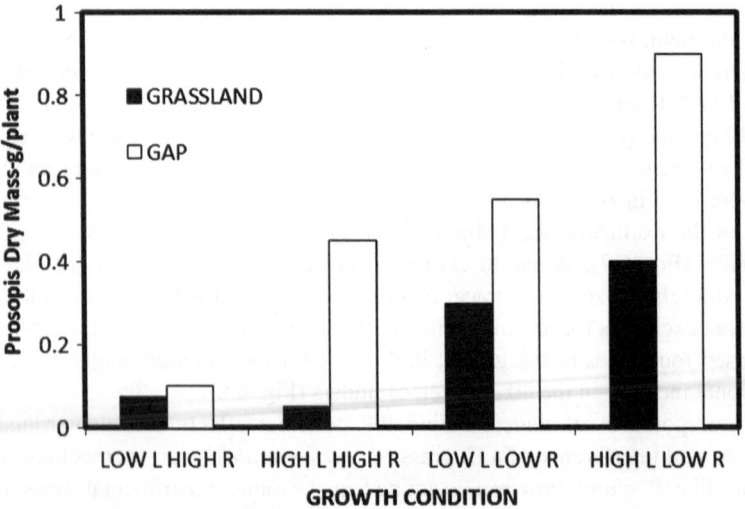

Fig. 6.9 Total *Prosopis glandulosa* dry mass was harvested from four different conditions or treatments. The woody plant was grown in an experimental (planted) *B. curtipendula* grassland or gaps (no grass plants), where light levels were adjusted to low light (50 % *shading*) or high light (*no shading*) treatments and with high belowground grass dry mass (2 cm root excluders) or low grass root mass (40 cm root excluders). Two-way ANOVA for dry mass was significant ($P<0.05$) with a significant location by treatment interaction ($P<0.0001$)

logarithmic relationship was found ($y = -0.27\ln(x) + 0.612$, $R^2 = 0.75$, $P = 0.0039$). As *B. curtipendula* belowground or root dry mass increased, total *P. glandulosa* dry mass decreased until it reached approximately zero at a *B. curtipendula* belowground or root dry mass of approximately 36 g/100 cm^2.

There are additional field and greenhouse experiments that should be completed to further our understanding of the interactions of both C_3 and C_4 grasses and the woody legumes and other woody plants that are and have been encroaching into arid and semiarid tropical and subtropical grasslands around the world. Is competition important in the encroachment or invasion of these grasslands? Yes, it is important, but it is not the only factor that should be considered. The grasses suppress or reduce the growth of the woody legumes and other encroaching species. We have seen this in all the experiments that we have completed. However, there were no woody legume mortalities. Another factor is important in preventing encroachment of woody plants, including the woody legumes into the arid and semiarid grasslands. That factor seems to be fire, both intensity and frequency.

References

Begon M, Townsend CR, Harper JL (2006) Ecology: from individuals to ecosystems, 4th edn. Blackwell, Malden, MA

Bush JK, Van Auken OW (1989) Soil resource levels and competition between a woody and herbaceous species. Bull Torrey Bot Club 116:22–30

Bush JK, Van Auken OW (1991) Importance of time of germination and soil depth on growth of *Prosopis-glandulosa* (leguminosae) seedlings in the presence of a C-4 grass. Am J Bot 78:1732–1739

Bush JK, Van Auken OW (1995) Woody plant-growth related to planting time and clipping of a C-4 grass. Ecology 76:1603–1609

Cohn EJ, Van Auken OW, Bush JK (1989) Competitive interactions between *Cynodon-dactylon* and *Acacia-smallii* seedlings at different nutrient levels. Am Midl Nat 121:265–272

Correll DS, Johnston MC (1979) Manual of the vascular plants of Texas. The University of Texas at Dallas, Richardson, TX

Gould FW (1969) Texas plants: a checklist and ecological summary. Texas Agricultural Experiment Station Bull, College Station, TX

Grace JB (1995) On the measurement of plant competition intensity. Ecology 76:305–308

Grace JB, Tilman D (1990) Perspectives on plant competition. Academic, New York

Grime JP (1979) Plant strategies and vegetation processes. Wiley, New York

Harper JL (1977) Population biology of plants. Academic, New York

Taylor FB, Hailey RB, Richmond DL (1966) Soil survey of Bexar County, Texas. USDA, Soil Conservation Service, Washington

Twolan-Strutt L, Keddy PA (1996) Above- and belowground competitive intensity in two contrasting wetland plant communities. Ecology 77:256–270

Van Auken OW (1994) Changes in competition between a C_4 grass and a woody legume with differential herbivory. Southwestern Nat 39:114–121

Van Auken OW (2000) Shrub invasions of North American semiarid grasslands. Annu Rev Ecol Syst 31:197–215

Van Auken OW (2009) Causes and consequences of woody plant encroachment into western North American Grasslands. J Environ Manage 90:2931–2942

Van Auken OW, Bush JK (1987) Interspecific competition between *Prosopis-glandulosa* torr (honey mesquite) and *Diospyros-texana* scheele (texas persimmon). Am Midl Nat 118:385–392

Van Auken OW, Bush JK (1988) Competition between *Schizachyrium scoparium* and *Prosopis glandulosa*. Am J Bot 75:782–789

Van Auken OW, Bush JK (1989) *Prosopis-glandulosa* growth – influence of nutrients and simulated grazing of *Bouteloua-curtipendula*. Ecology 70:512–516

Van Auken OW, Bush JK (1990) Importance of grass density and time of planting on *Prosopis-glandulosa* seedling growth. Southwestern Nat 35:411–415

Van Auken OW, Bush JK (1997) Growth of *Prosopis glandulosa* in response to changes in aboveground and belowground interference. Ecology 78:1222–1229

Chapter 7
Global Climate Change

The climate of the Earth has been changing ever since the Earth has had an atmosphere. Sometimes the changes are dramatic over relatively short time intervals, and other times the changes have been minor with long periods of relative stability. These global climate changes continue today. It is warmer today than it has been in the recent past, and the Earth is expected to get even warmer. The cause of the recent rise in temperature is the elevated level of CO_2 in the atmosphere (Mackenzie 2003). The expected increased warming of the Earth is not a new phenomenon, it has happened before. An example occurred approximately 56 million years ago, at the end of the Paleocene and the beginning of the Eocene. The Earth was much warmer compared to today, and it was about to get warmer during this Paleocene–Eocene thermal maximum (Kunzig and Block 2011). The warming event was triggered by a massive release of carbon, probably CO_2 or methane. The duration of high atmospheric carbon levels and high temperatures at that time is thought to be about 150,000 years, the time it took to reabsorb the carbon that had been released into the atmosphere. The amount of carbon released into the atmosphere at that time is estimated to be about as much as could be released in the near future if all of the stored fossil fuel was burned by man.

More recently, during the past 2 million years, throughout the Pleistocene, the climate of the Earth was mostly cooler than today (McDowell et al. 1995; Van Devender 1995; Martin 1999). There were probably 15–20 glacial periods during that time with associated interglacials (Imbrie and Imbrie 1979). The interglacials were 10,000–20,000 years long and the glacials about ten times longer. The present warmer period or interglacial started about 12,500 years ago and is projected to last much longer than most of the previous interglacials, perhaps more than 50,000 years (Berger and Loutre 2002). As temperatures warmed during the interglacials, the glaciers melted and plant communities around the world moved, adjusted, and changed their distributions. The reverse would be true at the start of the cooling that would precede the glacial periods.

O.W. Van Auken and J.K. Bush, *Invasion of Woody Legumes*, SpringerBriefs in Ecology 4,   43
DOI 10.1007/978-1-4614-7199-8_7, © Springer Science+Business Media New York 2013

During preceding glacial and interglacial periods, plant communities moved as the climate warmed or cooled without the influence of man or his animals. Today anthropogenic forces, man and his animals, play a major role in the development and distribution of plants and their communities. However, tying together causes and effects of these changes is much harder to do. Changes in distributions of former plant communities have been noted, and locations have been estimated. Many of these modifications in distributions of populations of plants and animals in the American southwest and other places have been estimated from pollen records and fossil packrat middens (Betancourt et al. 1990; Miller and Wigand 1994; Van Devender 1995; Martin 1999). In addition, the use of radiocarbon dating and oxygen isotope data from deep sea cores has been interpreted and used to develop a single time series to represent global changes in this and other areas throughout the world (McDowell et al. 1995).

Widespread shifts in plant community distributions are predicted in response to future altered rainfall and temperature regimes associated with the changing global climate (Adams et al. 2009; Kane et al. 2011). Projected higher temperatures are expected to compound the effects of severe droughts and elicit shifts in woody plant mortality and thus plant communities. The most dramatic shifts are expected to be caused by hurricanes, tornados, freezes, fires, and extreme droughts (Breshears et al. 2005). These predicted changes in woody plant mortality could change a plant community dramatically. Forests or woodlands could be converted to grasslands or savannas. This could be considered the reverse or a reset of community succession or a restart of the successional process (Begon et al. 2006). If the disturbance was a severe drought and caused the death of the majority of the dominant woody plants as occur recently in the American southwest, the resulting community could be converted to grassland. This could be the start of another successional process or event. The grassland could be encroached by various woody plants including woody legumes. With the encroachment, establishment, survival, and growth of the woody legumes and possibly other woody plants, a savanna would ensue, followed by community succession leading to the area climax community. The climax community would be determined by the interaction of all of the environmental conditions that created the pre-disturbance community. All of these factors have to be taken into consideration or examined for the long term, not the short term. Short-term interpretation of long-term events lead to mistakes, misinterpretations, and misunderstandings.

Prior to the current warming cycle, pine parkland and juniper woodland or savanna was common in what are now southwestern North American semiarid grasslands or desert grasslands (Betancourt et al. 1990; Miller and Wigand 1994; Van Devender 1995; Martin 1999). These communities have appeared and disappeared in these same areas over the millennia as atmospheric CO_2 levels and global temperatures have changed. The semiarid and desert grasslands of the American southwest were lower in elevation and more southward in the past. Desert shrublands were lower in elevation and more southward as well. In spite of what is known of the locations of these communities, it has been difficult to link climatic changes over the past 160 years to recent vegetation changes because of the relatively short time, considerable variability, and various interacting factors.

Evidence demonstrating links in precipitation patterns or temperature to recent shrub or woody plant encroachment in southwestern North America since the 1870s are tenuous. However, a recent prolonged drought has caused considerable Pinion and *Juniperus* mortality in Pinion-Juniper woodlands in this area (Breshears et al. 2005; Kane et al. 2011). Similar mortalities in other communities have not been documented. The severity and frequency of similar droughts are expected to increase as the climate continues to be modified. Nevertheless, the irregular nature of encroachment of woody legumes like *Vachellia, Senegalia, Prosopis,* and others, with striking differences in density and basal area in contiguous, fenced, edaphically similar areas, would seem to rule out large-scale climatic influences as the major cause of the encroachment.

Increased levels of atmospheric CO_2 have been proposed as the cause of shrub and woody legume encroachment into grasslands throughout North America including the semiarid grasslands in the southwest (Mayeux et al. 1991; Idso 1992; Polley et al. 1992; Johnson et al. 1993). There is some evidence to link greater woody plant density to greater growth at higher atmospheric CO_2 levels (Morgan et al. 2007). This hypothesis is very attractive and could account for the synchronous, widespread encroachment of woody plants into semiarid grasslands and savannas throughout the world; however, proof is equivocal. Dramatic density differences in adjacent, fenced, edaphically similar areas would again seem to rule out large-scale climatic influences as the major cause of increased woody plant density and biomass (Archer 1994, 1995). The causes still seem to be more local and management related (Bahre and Shelton 1993; Van Auken 2009).

Regardless, the elevated CO_2 hypothesis is based on observations that the woody legumes and most other woody plants have the C_3 photosynthetic pathway and the grasses that are being replaced in southwestern grasslands have the C_4 photosynthetic pathway (Begon et al. 2006). The C_3 photosynthetic pathway is advantageous at higher levels of CO_2. However, there are difficulties with this hypothesis (Archer et al. 1995). Many C_3 and C_4 species have similar photosynthetic characteristics. There are a number of C_4 grasses that are more responsive to augmented levels of CO_2 than previously thought. Substitution or replacement of various C_3 grasses in the northern cold deserts or cool temperate grasslands by encroachment of C_3 woody shrubs is not explained by the elevated CO_2 hypothesis. In southwestern North American, C_3 woody plants like *Vachellia, Senegalia,* and *Prosopis* are replacing the C_4 grasses, but C_3 grasses are not replacing C_4 grasses. Once more, in areas with similar soils, fences preventing the movement of domestic herbivores reduce the encroachment of these C_3 woody shrubs and small trees.

Finally, if one examines the timing when most of the encroachment occurred, it does not match the time of the highest levels of atmospheric CO_2. There is a temporal disparity between these two factors. The highest levels of atmospheric CO_2 followed the greatest extent of woody plant encroachment. Populations of woody plants, especially the woody legumes and grasses, have shifted during the Holocene, but the shifts do not seem to be explained by elevated levels of atmospheric CO_2. Furthermore, not all elevated CO_2 studies have demonstrated a fertilizer or stimulatory effect, suggesting other limitations or constraints on the grass plants that are being replaced. These could be nitrogen, phosphate, water, a combination of these, or possibly other factors. Therefore, the CO_2 enrichment hypothesis does not seem to explain the encroachment of woody legumes or other woody plants into the southwestern semiarid grasslands or other grasslands. Another factor or factors seem to be suggested.

References

Adams HD et al (2009) Temperature sensitivity of drought-induced tree mortality portends increased regional die-off under global-change-type drought. Proc Natl Acad Sci 106:7063–7066

Archer S (1994) Woody plant encroachment into southwestern grasslands and savannas: rates, patterns and proximate causes. In: Vavra M, Laycock WA, Pieper RD (eds) Ecological implications of livestock herbivory in the West. Society for Range Management, Denver, pp 13–69

Archer S (1995) Tree-grass dynamics in a *Prosopis*-thronscrub savanna parkland: reconstructing the past and predicting the future. Ecoscience 2:83–99

Archer SR, Schimel DS, Holland EH (1995) Mechanisms of shrubland expansion: land use, climate or CO_2. Climate Change 29:91–99

Bahre CJ, Shelton ML (1993) Historic vegetation change, mesquite increases, and climate in southeastern Arizona. J Biogeogr 20:489–504

Begon M, Townsend CR, Harper JL (2006) Ecology: from individuals to ecosystems, 4th edn. Blackwell, Malden, MA

Berger A, Loutre MF (2002) An exceptionally long interglacial ahead? Science 297:1287–1288

Betancourt JL, Van Devender TR, Martin PS (1990) Synthesis and prospectus. In: Betancourt JL, Van Devender TR, Martin PS (eds) Packrat Middens: The last 40,000 years of biotic change. University of Arizona Press, Tucson, pp 435–447

Breshears DD, Cobb NS, Rich PM, Price KP, Allen CD (2005) Regional vegetation die-off in response to global-change type drought. Proc Natl Acad Sci USA 102:15144–15148

Idso SB (1992) Shrubland expansion in the American southwest. Climate Change 22:85–86

Imbrie J, Imbrie KP (1979) Ice ages: solving the mystery. Enslow, Short Hills, NJ

Johnson HB, Polley HW, Mayeux HS (1993) Increasing CO_2 and plant-plant interactions: effects on natural vegetation. Vegetatio 104–105:157–170

Kane JM, Meinhardt KA, Chang T, Cardall BL, Michalet R, Whitham TG (2011) Drought-induced mortality of a foundation species (Juniperus monosperma) promotes positive afterlife effects in understory vegetation. Plant Ecol 212:733–741

Kunzig R, Block I (2011) World without ice. Natl Geographic 10:90–109

Mackenzie FT (2003) Our changing planet: an introduction to Earth system science and global environmental change. Prentice Hall-Pearson Education, Upper Saddle River, NJ

Martin PS (1999) Deep history and a wilder West. In: Robichaux RH (ed) Ecology of the Sonoran Desert plants and plant communities. The University of Arizona Press, Tucson, pp 255–290

Mayeux HS, Johnson HB, Polley HW (1991) Global change and vegetation dynamics. In: James LF, Evans JO, Ralphs MH, Sigler BJ (eds) Noxious range weeds. Westview Press, Boulder, Colorado, pp 62–74

McDowell PF, Webb T III, Bartlein PJ (1995) Long-term environmental change. In: Powell TM, Steele JH (eds) Ecological time series. Chapman & Hall, New York, pp 327–370

Miller RF, Wigand PE (1994) Holocene changes in semiarid pinyon-juniper woodlands: response to climate, fire, and human activities in the Great Basin. Bioscience 44:465–474

Morgan JA, Milchunas DG, LeCain DR, West M, Mosier AR (2007) Carbon dioxide enrichment alters plant community structure and accelerates shrub growth in the shortgrass steppe. Proc Natl Acad Sci USA 104:14724–14729

Polley HW, Johnson HB, Mayeux HS (1992) Carbon dioxide and water fluxes of C_3 annuals and C_3 and C_4 perennials at subambient CO_2 concentrations. Funct Ecol 6:693–703

Van Auken OW (2009) Causes and consequences of woody plant encroachment into western North American Grasslands. J Environ Manage 90:2931–2942

Van Devender TR (1995) Desert grassland history: changing climates, evolution, biography, and community dynamics. In: McClaran MP, Van Devender TR (eds) The desert grassland. University of Arizona Press, Tucson, pp 68–99

Chapter 8
Management and Community Restoration

Periodic fires in grasslands, savannas, and encroached woodlands are very important and should not be overlooked or underestimated as management tools. Without fire, which was a natural part of the grassland ecosystems (Fig. 8.1), the woody plants are favored rather than the grasses (Van Auken 2009). An excellent visual example of the result of using fire to control the encroachment of woody species in tall grass prairie is from the Konza prairie in Kansas (Fig. 8.2). In the center of the photograph is a firebreak. To the left of the firebreak, the prairie was burned infrequently (once every 10 years). The encroachment of a variety of woody plants is easily seen. To the right of the firebreak, the prairie was burned frequently (once every 2 years). The lack of woody plants is obvious (this figure is modified from Van Auken (2009) and was taken by D. C. McKinley). However, burning discontinuous arid and semiarid shrublands or former grasslands today is difficult because of the patchy nature of the communities and the limited amount of light, fluffy fuel (Schlesinger et al. 1996; Jurena and Van Auken 1998).

Should managing or attempting to manage the encroachment of woody legumes and other woody plants into former grasslands be about the grasslands or about the new communities that are created, the savannas, or woody legume communities? In the past, the main concern seems to have been about the grasslands or former grasslands that have been encroached. There have been significant reductions in biodiversity as woody plants have increased their density and basal area in these former grasslands (Ratajczak et al. 2012). A bigger concern seems to be the loss of herbaceous or grass productivity. The density, cover, and biomass of certain woody species, especially the woody legumes, have increased as the density, cover, and biomass of many species of C_3 and C_4 grasses and other herbaceous species have decreased in the vast area of grassland where encroachment has been occurring (Knapp et al. 2008a, b). The woody legumes are not the only encroaching species. Various species in more than 30 genera of woody plants have been reported to be encroaching into worldwide grasslands (Archer et al. 1995). However, the woody legumes seem to be principle encroachers.

O.W. Van Auken and J.K. Bush, *Invasion of Woody Legumes*, SpringerBriefs in Ecology 4,   49
DOI 10.1007/978-1-4614-7199-8_8, © Springer Science+Business Media New York 2013

Fig. 8.1 Picture in upper panel is a native tall grass prairie in North Texas in early fall before a controlled burn, and the *lower panel* is of a controlled burn in the same tall grass prairie. The fire was a head fire and covered approximately 25 ha. All woody plants in the burn area were top killed. Photographs were taken by the senior author

Major shifts in allocation of biomass, carbon, and nitrogen pools have been reported from largely belowground in grasslands to aboveground in developing woodlands (McKinley et al. 2008b). These shifts have been reported from *Juniperus* woodlands recently converted from grasslands, not legume woodlands, but the same types of changes appear to be occurring in the legume woodlands, perhaps with the exception of increased nitrogen fixation in the legume woodlands (Bush 2008). Continuing changes to other soil components and soil processes are not as clear. Late in encroachment, distribution of soil nutrients is less uniform and more

Fig. 8.2 Photograph demonstrates the management potential of using fire in grassland ecosystems to control or reduce the encroachment and establishment of woody plants. Picture was taken in the direction of a firebreak along an access road which is near the center of the photograph. To the *left* of the firebreak, the prairie was burned infrequently (once every 10 years). The encroachment of a variety of woody plants is easily seen. To the *right* of the firebreak, the prairie was burned frequently (once every 2 years). The lack of woody plants is obvious (this figure is modified from Van Auken (2009) and was taken by McKinley DC)

variable compared to former grasslands (Schlessinger et al. 1990; Jurena and Van Auken 1998; Schlesinger et al. 1999; Throop and Archer 2008). Nutrient and organic concentrations are higher below the woody plant or shrub canopy rather than the interspaces. Soil nitrogen levels are higher below the canopy of encroaching woody legumes such as *Prosopis, Senegalia,* and *Vachellia* (Bush 2008; Jurena and Van Auken 1998).

With true invasive species, pools of biogeochemicals and material fluxes have changed, but there is considerable variability (Ehrenfeld 2010) and considerable effort to prevent the establishment of the invaders. Where various other woody species are encroaching or have encroached into grasslands, changes in soil resources are just being revealed. In xeric sites, encroachment caused a decrease in aboveground net primary production, but in more mesic sites, there was an increase in production because of high woody plant leaf area (Knapp et al. 2008b) which is woody plant production. In central North American *Juniperus* woodlands, carbon and nitrogen increased in woody plant biomass, but there was little change in soil availability of nitrogen (McKinley and Blair 2008). There were small changes in nitrogen cycle processes in these same woodlands that were recently converted from grasslands (McKinley et al. 2008a).

Demonstrated biggest changes in the location of the biomass from belowground in the grassland to aboveground in the woodland and the form of the biomass from herbaceous to woody are very important to most of the herbivorous animals. Herbaceous biomass supports the grazers, while the woody biomass supports some browsers. Because of the shift in plant biomass, these former grassland communities can no longer support large populations of grazers and a pastoral economy (Campbell et al. 1997; Reynolds et al. 2007). Management has changed in many southwestern North American arid and semiarid grasslands from cattle and other grazing animals to white-tailed deer and other browsing ungulates (Doughty 1983).

However, there is some evidence that shrub or woody plant cover has started to stabilize, with little continued temporal expansion (Browning et al. 2008). The encroachment of some of these woody plants into grasslands appears to have slowed (Miller et al. 2005), suggesting that the period of greatest encroachment may be past but evidence is limited. Perhaps the majority of grasslands that could be encroached have been encroached. However, there are many ongoing encroachment studies.

Past woody plant or brush encroachment was treated as a problem associated with the encroaching woody species. What was not recognized was the effect of constant, long-term, high levels of herbivory by domestic herbivores on the grass plants. Grass plants can tolerate some herbivory but not a lot. With the loss of aboveground parts due to herbivory, roots and other belowground parts are reduced, and the ability of the grasses to obtain soil resources including water is reduced. This is followed by a reduction in aboveground carbon fixation and production (Van Auken and Bush 1989, 1997; Bush and Van Auken 1995). Encroaching woody species were considered aggressors, aliens, and invaders. Using anthropogenic terms suggested, these species were special. In fact they were growing extremely well in disturbances or gaps without competition and without fire. These are areas or conditions where these species normally grow very well with the niche requirements needed for their typical growth. The biology and ecology of what was happening was not considered or it was ignored. What was happening to the grass in these grassland or rangeland was not deemed important except that it appeared to be declining and the woody plants were the cause. This decline appeared to be very important to ranchers and cattle or domestic animal production. Thus, grassland and rangeland managers used various techniques including both mechanical and chemical to try unsuccessfully to control the perceived problem (Scifres 1980; Heitschmidt and Struth 1991; Taylor 2008; Taylor et al. 2012).

Unfortunately, the mature woody legumes or other mature woody plants were targeted, not the newly germinated or the newly emerged woody plant seedlings (Fig. 8.3). Healthy grass competitors can reduce and slow the growth of the woody legumes, but they do not seem to be able to totally prevent their growth (Fig. 8.4). Combinations of burning and biological control were tried without a lot of apparent success, probably because of little light, fluffy fuel remaining after years of heavy grazing. Woody plant mortality is size and fire dependent, and large plants in these encroached grasslands are fire resistant (Collins and Wallace 1990). In these managed, manipulated communities, the domestic animals were usually put back on the treated area as soon as the grasses started to regrow after mechanical, chemical, or other treatment. This was usually after one or more rains. This type of treatment would reset the grassland-woodland succession with an approximate 20-year cycle of treatment, grass regrowth, and woody plant encroachment followed by re-treatment. This kind of procedure or treatment is very expensive, not very practical, and really not ecologically sound.

Fig. 8.3 The *left panel* is a mature *Prosopis glandulosa* tree with a grassland encroached with mature *Prosopis* plants in the background. Tree-sized plants are difficult to kill with fire or herbicide. If the trees were cut down and used for firewood, stumps would re-sprout and continue to grow. The *right panel* is a recently germinated *P. glandulosa* seedling (below the penny, *white arrow*) that could be easily killed by fire or eaten and killed by various small or relatively small herbivores. The photographs were taken by the senior author

Fig. 8.4 The photograph shows a comparison of two pots with *Prosopis glandulosa* and *Cynodon dactylon,* an introduced C_4 grass growing together. The pots are 15 cm deep and all of the plants were grown together for 12 weeks in competition. There is one *P. glandulosa* plant in each pot (note the tip of the *white arrow*). *Cynodon dactylon* density was two plants per pot on the *left* and 32 plants per pot on the *right*. The photograph was taken by the junior author

Various combinations of treatments have been used with slightly greater success, but considerable woody plant regrowth still occurred and expenses were high. Without fires, treatments are usually expensive with limited success. Management of heavily encroached grasslands or savannas usually includes some degree of mechanical or chemical treatment, but success is short term and is still limited.

Fig. 8.5 Three examples of unburned grassland recently encroached by *Juniperus* (*top left*), *Opuntia* (*top right*), and *Prosopis* and a different species of *Opuntia* (*bottom*). Photographs were taken by the senior author

Reversing the encroachment or succession process or going from a woodland or shrubland to grassland is complex and difficult (Ansley and Wiedemann 2008). Potential biotic controls are usually more sophisticated, may be combined with fire or mechanical treatment, and may include genetic manipulation of some of the browsing species (Taylor 2008; Taylor et al. 2012).

If intermittent or periodic fires are not allowed or do not occur in these grasslands communities, they will be converted to savannas, shrublands, or woodlands (Fig. 8.5). The frequency of fire in native grasslands in the past was and still is quite variable, but there seems to be general agreement that recurring fires are required to control or reduce the encroachment, establishment, density, and growth of woody plants in most if not all grasslands (Collins and Wallace 1990; Scholes and Walker 1993; Simmons et al. 2007). The use of various types of models have produced similar findings, heavy grazing or low fire frequency results in encroachment of woody plants which is followed by increased woody plant density (Fuhlendorf et al. 2008). Fire results are not simple, and there are a number of potential environmental and ecological interactions that could occur between fires and topography, soil type, temperature, rainfall, number and kind of herbivores, and amount of light, fluffy fuel. These factors will determine the nature, density, and location of woody plants in a given landscape (Humphrey 1958; Collins and Wallace 1990; McPherson 1995).

Of course, fire frequency and intensity are linked to climatic patterns and conditions including rainfall amount and pattern as well as temperatures that are linked to global phenomena (Swetnam and Betancourt 1990). In southwestern North America, large fires usually high elevation fires occur after dry springs with smaller fires following wet springs with lower elevation semiarid grasslands probably following the same pattern.

Evidence of past grassland fires throughout the world is sparse, difficult to find, and not from the usual scientific sources. Some historical evidence for fires in southwestern North America comes from newspaper reports, from the earliest travelers, and from early settlers (Humphrey 1958; Inglis 1964; Bahre and Shelton 1993). Although evidence concerning past fires is tenuous, evidence suggests grasslands require fires, but some do not agree.

Encroachment of grasslands by woody legumes and other woody species caused by man and his grazing animals has probably been going on in many parts of the world for thousands of years. However, in southwestern North America, the process has been only going on a relatively short time, probably for a few hundred years. Apparently encroachment and changes in the composition and structure of arid and semiarid low elevation grasslands in many parts of southwestern North America seemed to have started after the beginning of large-scale cattle ranching and fire exclusion in the 1850s–1870s (Humphrey 1958; Bahre 1991, 1995; Neff et al. 2008). Unfortunately, fires in these low elevation communities today are rare because of changes coupled to high intensity, continuous grazing that reduces the amount of light, fluffy fuel required for fires. Today the presence of woody plants, little fine fuel, and community fragmentation are causes of reduced fire frequency (Collins and Wallace 1990). Grass biomass and all fire characteristics have declined with the increase in density and size of woody plants in these communities.

Seedlings of woody plants found in grasslands are sensitive to fire (McPherson 1995). Plant sensitivity is size dependent and mortality is mostly related to small size. Small woody plants are killed by fire and larger ones are top killed and some will not re-sprout if top killed. Woody plants are susceptible to fire mortality until they are fairly large and have a thick bark. Additionally, if encroaching woody plants do not produce seeds before the next fire, or do not re-sprout after being top killed by fire, these encroached grasslands would remain relatively free of woody plants. Fire-tolerant woody species would be top killed but suppressed by reoccurring fires and remain in the grassland, but at a small size (Archer 1994). With a reduction of the grassland fuel by herbivory, fire frequencies would decrease and approach zero, promoting encroachment, establishment, growth, and increased woody legume and other woody plant cover and density.

Management of southwestern North American grasslands and grasslands throughout the world is a difficult task. Encroachment of woody legumes and other woody species will continue with fire suppression. The goal of management of these grasslands should be sustainability. That should include sustainable grass production that leads to sustainable animal production. That does not mean all grass can be harvested by the animals. In order to maintain grasslands as grasslands, they should be burned at reasonable intervals to suppress or remove the woody plants, and the intervals would probably depend on the rainfall. With a reasonable schedule of burning and reduced levels of herbivory, grasslands could remain as grasslands and animal productivity could be sustained. The level of herbivory would certainly have to be adjusted to area conditions especially rainfall.

Can woody legume and other woody plant savannas or woodlands be maintained or restored to grasslands? This is a difficult question without easy answers. Yes it can probably be done, but not cheaply. Two hundred to 300 or more years of excessive grazing and harvest cannot be overcome with one fire followed by a return to a previous ineffective management system. Recovery or return of a savanna or woodland to a grassland state will not be easy. The "spray and pray" philosophy of the past did not and will not work.

The future of these woody legume communities or former grasslands is difficult to predict, especially in our changing world. Secondary succession is occurring or has occurred, and the communities will be different in the world of the future. These communities are very dynamic and they are also economically, biologically, and ecologically necessary and needed. Knowledge gained from observations and experiments reported here and other places are critical in defining observations and experiments needed in the future. The track through the future is not clear, but each new discovery helps light the way.

References

Ansley RJ, Wiedemann HT (2008) Reversing the woodland steady state: vegetation responses during restoration of *Juniperus*-dominated grasslands with chaining and fire. In: Van Auken OW (ed) Western North American *Juniperus* communities: a dynamic vegetation type, vol 196. Springer, New York, pp 272–290

Archer S (1994) Woody plant encroachment into southwestern grasslands and savannas: rates, patterns and proximate causes. In: Vavra M, Laycock WA, Pieper RD (eds) Ecological implications of livestock herbivory in the west. Society for Range Management, Denver, pp 13–69

Archer SR, Schimel DS, Holland EH (1995) Mechanisms of shrubland expansion: land use, climate or CO_2. Climate Change 29:91–99

Bahre CJ (1991) A legacy of change: historic human impact on vegetation on the Arizona borderlands. University of Arizona Press, Tucson

Bahre CJ (1995) Human impacts on the grasslands of southeastern Arizona. In: McClaran MP, Van Devender TR (eds) The desert grassland. The University of Arizona Press, Tucson, pp 230–264

Bahre CJ, Shelton ML (1993) Historic vegetation change, mesquite increases, and climate in southeastern Arizona. J Biogeogr 20:489–504

Browning DM, Archer SR, Asner GP, McClaran MP, Wessman CA (2008) Woody plants in grasslands: post-encroachment stand dynamics. Ecol Appl 18:828–944

Bush JK (2008) Soil nitrogen and carbon after twenty years of riparian forest development. Soil Sci Soc Am J 72:815–822

Bush JK, Van Auken OW (1995) Woody plant-growth related to planting time and clipping of a C-4 grass. Ecology 76:1603–1609

Campbell BD, Stafford-Smith DM, McKeon GM (1997) Elevated CO_2 and water supply interactions in grasslands: a pastures and rangelands management perspective. Global Change Biol 3:177–187

Collins SL, Wallace LL (1990) Fire in North American tallgrass prairie. University of Oklahoma Press, Norman, Oklahoma

Doughty RW (1983) Wildlife and man in Texas: environmental change and conservation. Texas A & M University Press, College Station, TX

Ehrenfeld JG (2010) Ecosystem consequences of biological invasions. Annu Rev Ecol Evol Syst 41:59–80

Fuhlendorf SD, Archer SR, Smeins FE, Engle DM, Taylor CA Jr (2008) The combined influence of grazing, fire, and herbaceous productivity on tree-grass interactions. In: Van Auken OW (ed) Western North American *Juniperus* communities: a dynamic vegetation type, vol 196. Springer, New York, pp 219–238

Heitschmidt RK, Struth JW (1991) Grazing management: an ecological perspective. Timberline Press, Portland

Humphrey RR (1958) The desert grassland: a history of vegetational change and an analysis of causes. Bot Rev 24:193–252

Inglis JM (1964) A history of vegetation on the Rio Grande plain. Texas Parks and Wildlife Department. Texas Parks and Wildlife Department, Austin, Texas

Jurena PN, Van Auken OW (1998) Woody plant recruitment under canopies of two acacias in a southwestern Texas shrubland. Southwestern Nat 43:195–203

Knapp AK, Briggs JM, Collins SL, Archer SR (2008a) Shrub encroachment in North American grasslands: shifts in growth form dominance rapidly alters control of ecosystem carbon inputs. Glob Change Biol 14:615–623

Knapp AK, McCarron JK, Silletti GA, Hoch GL, Heisler MS, Lett JM, Blair JM et al (2008b) Ecological consequences of the replacement of native grassland by *Juniperus virginiana* and other woody plants. In: Van Auken OW (ed) Western North American *Juniperus* communities: a dynamic vegetation type, vol 196. Springer, New York, pp 156–169

McKinley DC, Blair JM (2008) Woody plant encroachment by *Juniperus virginiana* in a mesic native grassland promotes rapid carbon and nitrogen accrual. Ecosystems 11:454–468

McKinley DC, Rice CW, Blair JM (2008a) Conversion of grassland to coniferous woodland has limited effects on soil nitrogen cycle processes. Soil Biol Biochem 40:2627–2633

McKinley DC, Morris MD, Blair JM, Johnson LC (2008b) Altered ecosystem processes as a consequence of *Juniperus virginiana* L. encroachment into North American tallgrass prairie. In: Van Auken OW (ed) Western North American *Juniperus* communities: A dynamic vegetation type, vol 196. Springer, New York, pp 170–187

McPherson GR (1995) The role of fire in the desert grassland. In: McClaran MP, Van Devender TR (eds) The desert grassland. The University of Arizona Press, Tucson, pp 131–151

Miller RF, Bates JD, Svejcar TJ, Pierson FB, Eddleman LE (2005) Biology, ecology and management of western juniper (*Juniperus occidentalis*). Agriculture Experiment Station, Oregon State University, Corvalis, Oregon

Neff JC, Ballantyne AP, Farmer GL, Mahowald NM, Conroy JL, Landry CC, Overpeck JT et al (2008) Increasing eolian dust deposition in the western United States linked to human activity. Nat Geosci 1:189–195

Ratajczak Z, Nippert JB, Collins SL (2012) Woody encroachment decreases diversity across North American grasslands and savannas. Ecology 93:697–703

Reynolds JF, Smith DMS, Lambin EF, Turner BL II (2007) Global desertification: building a science for dryland development. Science 316:847–851

Schlesinger WH, Raikes JA, Hartley AE, Cross AF (1996) On the spatial pattern of soil nutrients in desert ecosystems. Ecology 77:364–374

Schlesinger WH, Abrahams AD, Parsons AJ, Wainwright J (1999) Nutrient losses in runoff from grassland and shrubland habitats in southern New Mexico: rainfall simulation experiments. Biogeochemistry 45:21–34

Schlessinger WH, Reynolds JF, Cunningham GL, Huenneke LF, Jarrell WH, Virginia RA, Whitford WG (1990) Biological feedbacks in global desertification. Science 247:1043–1048

Scholes RJ, Walker BH (1993) An African Savanna: synthesis of the Nylsvley study. Cambridge University Press, Cambridge

Scifres CJ (1980) Brush management principles and practices for Texas and the Southwest. Texas A&M University Press, College Station, TX

Simmons MT, Archer SR, Ansley RJ, Teague WR (2007) Grass effects on tree (*Prosopis glandulosa*) growth in a temperate savanna. J Arid Environ 69:212–227

Swetnam TW, Betancourt JL (1990) Fire-southern oscillation relations in the southwestern United States. Science 249:1017–1020

Taylor CA Jr (2008) Ecological consequences of using prescribed fire and herbivory to manage *Juniperus* encroachment. In: Van Auken OW (ed) Western North American *Juniperus* communities: a dynamic vegetation type, vol 196. Springer, New York, pp 239–252

Taylor CA Jr, Twidwell D, Garza NE, Rosser C, Hoffman JK, Brooks TD (2012) Long-term effects of fire, livestock herbivory removal and weather variability in Texas semiarid savanna. Rangeland Ecol Manage 65:21–30

Throop HL, Archer SR (2008) Shrub (*Prosopis velutina*) encroachment in a semi-desert grassland: spatial-temporal changes in soil organic carbon and nitrogen pools. Global Change Biol 14:1–12

Van Auken OW (2009) Causes and consequences of woody plant encroachment into western North American Grasslands. J Environ Manage 90:2931–2942

Van Auken OW, Bush JK (1989) *Prosopis glandulosa* growth – influence of nutrients and simulated grazing of *Bouteloua curtipendula*. Ecology 70:512–516

Van Auken OW, Bush JK (1997) Growth of *Prosopis glandulosa* in response to changes in aboveground and belowground interference. Ecology 78:1222–1229

Chapter 9
The Future

After examining what we have brought together here and what is presented in other places in the published literature, it appears that understanding woody legume communities, including savannas and woodlands, is a difficult task. The results of a number of experiments that are yet to be done are needed to fully understand these woody legume communities and what will happen to them in a warmer, high CO_2 world. There seem to be many physical, chemical, and biotic factors involved. It will probably be more difficult to manage these communities because many of these savannas and woodlands were most likely grasslands at some time in the relatively recent past. Consequently, understanding the interactions of the C_4 grasses and the various woody legumes should be added to the mix of what is required to understand these communities. Thus, the task is complicated, but it is even more difficult when relatively short-term temporal modifications (successional changes) and small-scale spatial variability are added as factors that may further obscure the understanding of long-term effects such as global climate transformations.

The invasion, or more correctly encroachment, of woody legumes into warm season grasslands has happened all over the world. The grasslands can be various types and seem to range from having moderate rainfall to dry-semiarid and arid; thus, there seems to be a rainfall gradient involved. This process involving the woody legumes really seems to be secondary succession, but not quite the classical eastern North American farmland abandonment start of secondary succession. Farmland abandonment could be the start and is in some cases, but usually one or another species of woody legume encroaches (establishes) in the grassland and then expands its population because of heavy and constant grazing and lack of fire. The environmental conditions present in past grasslands have changed; consequently present-day grasslands are very different from what they were, and future grasslands will continue to adjust to the new conditions. A grassland community goes through secondary succession after heavily grazing and no fire. Changes seem to be from grassland, to shrub land, to legume savanna, to legume woodland, and then to possibly another type of nonlegume woodland or some other type of nonlegume community. The last or later community in many cases seems undefined or unknown

O.W. Van Auken and J.K. Bush, *Invasion of Woody Legumes*, SpringerBriefs in Ecology 4, 59
DOI 10.1007/978-1-4614-7199-8_9, © Springer Science+Business Media New York 2013

at this time and will probably be determined by future conditions, especially amount and intensity of grazing, fire, and certainly the amount of rainfall or drought.

Apparently these C_4 warm season grasslands and woody legume communities will not remain the same in the future unless management is modified. Grazing intensity by domestic animals and fire or burning frequencies should be adjusted to at least moderate levels. Grazing and fire frequency as well as intensity are factors of paramount importance to understanding woody legume encroachment and the functioning of these communities. These factors are all interrelated and interdependent. The intensity and frequency of fire depends on many factors including grazing intensity, precipitation, temperature, soil depth, soil nutrient levels, and probably others that are not known for many woody legume communities. If grasslands are not burned, ecological succession will change. The grasslands will not remain the same in time, and woody species such as the various woody legumes and a few other woody species will encroach into the grasslands, and the grasslands will become woody legume savannas, and savannas will become woodlands that will continue to change.

Fire and its effects on most of these legume communities have not been carefully examined. This is especially true for the *Vachellia* and *Senegalia* communities in the Americas. Fire has been used in attempts to control *Prosopis* encroachment, but usually late in the encroachment process, and required conditions and results are not yet well defined. What happens if the legume communities are not burned? Results seem varied. In more mesic areas, succession continues and the woodlands are replaced in time with species more tolerant of shaded understory conditions, and probably requiring higher levels of soil nitrogen or possibly other undefined conditions. This woodland successional sequence has been demonstrated for two woody legume species in south central North America, and a similar sequence has been demonstrated in southern Africa. However, in the grasslands, legume savannas, and woodlands of the southern Great Plains of central North America including north Texas, the conclusions are not as clear. The same is true for some of the arid and semiarid grassland biomes and former grasslands of southwestern North America. It is not clear if other drought-tolerant and shade-tolerant species will replace the various woody legumes in these habitats. The answer to this question remains unrequited at this time. Have the potential replacement species been identified? In a few cases they have been identified, but in most cases and for most conditions, they have not been identified.

The encroachment or invasion process really seems to be a modification of ecological succession known and studied for many years. However, in grasslands, the final stage in the succession has long been considered a mixture of various grasses and herbaceous plants, the grassland stage. The grassland stage has been considered limited and the "climax" community because of the amount of annual rainfall in the area of the community. This does not appear to be correct. In the past, the grassland communities seem to have been limited or controlled by the frequency and intensity of fire in what was called a "pyric climax." Today, grasslands are modified by the amount of herbivory and thus the amount of fuel for a fire. Could the succession in these grasslands or legume shrublands or woodlands be cyclic? Fire seems to reset the successional process. If the woodlands are burned, would or could these communities revert to open grasslands or savannas? This question is only partially answered and seems to depend on the stage of succession and fire frequency and intensity. What is the timeline of these changes? Also largely undefined at present and is dependent on local conditions that modify the rate of change.

Are domestic grazing animals important in this encroachment and succession of woody legumes? Yes, they are. Would the removal of domestic grazing animals prevent the encroachment of various woody legumes and other species into the arid and semiarid C_4 warm season grassland communities? Apparently the answer is no; the grasslands should be allowed to burn and then the woody species could be controlled or reduced. The removal of the domestic herbivores would modify the rate of change, but it would not prevent the woody legume plant encroachment. Grazing by domestic cattle would speed up the encroachment process, if the grazing was heavy and continuous. If grazing were at a lower level and frequency, the rate of the encroachment would be slower. The presence and competition of various C_3 and C_4 grasses with the woody legumes would slow down the encroachment process but not prevent it from occurring. Dead grass biomass is fuel for fires that can move through these biomes. If there was no grass biomass or it was reduced because of heavy grazing, there would be little or no fuel for the grassland fires. With enough fuel, the fires could be at a relatively high temperature and frequency, and the fires would top kill the woody legumes and reverse to some extent woody legume and other woody plant growth and succession. But, 200 years of heavy and almost continuous grazing would not be reversed with one fire.

Control of encroachment of woody legumes into these grassland communities must occur early in the encroachment or woody legume establishment process. It would probably be best if it occurred in the seedling stage or early growth stage of the encroaching plants. Control or management of woody legumes and other woody species in grasslands usually is attempted well after, usually years after, the woody plants germinate and establish. Thus, only top killing occurs and multiple fires are necessary to attempt density reduction or control. Seasonality of the fires is also an important potential effect on woody legume encroachment, but effects of seasonality of fires need to be examined more carefully, especially summer fires.

Large domestic grazing animals reduce the amount of aboveground grass biomass, and if grazing is continuous and intense, belowground grass biomass will be reduced as well. This would reduce the ability of the grass to regrow leaves and stems and belowground parts as well and therefore reduce the grasses ability to obtain belowground resources including water, nitrogen, phosphate, and other soil components. It would also reduce the grasses' ability to grow and compete in these ecosystems. Another thing that continuous and heavy grazing does is reduce the fuel available for grassland fires and the ability of a grassland community to carry a fire. So, any normal fire-caused woody plant mortality would cease.

What about the invertebrate herbivores in these grassland communities? Unfortunately, this topic has mostly been ignored. The upper layer of soil litter is where most of these invertebrates live, and this is where the woody legume seed germination, seedling emergence, and seedling establishing take place. If there is little or no litter or biomass from grass growth and decomposition, the soil surface will be essentially devoid of these invertebrate herbivores and the emerging woody legume seedlings would not be threatened by them. In the past, these soil or litter invertebrate herbivores could have been a major barrier to woody legume seedling establishment, but they have never been studied in this context.

Because of natural and anthropogenic factors, the story of woody legume encroachment into grasslands is more complicated. Natural atmospheric carbon dioxide levels of the Earth have increased over the past 12–15,000 years, and they are continuing to increase at rates modified by human activities. Increased atmospheric levels of CO_2 have aggravated the greenhouse effect, which has caused the mean annual temperature of the Earth to increase. Higher levels of atmospheric CO_2 will also stimulate plant carbon uptake or photosynthetic activity in many species. This could result in an increase in annual net primary production, as long as another required nutrient or factor such as soil nitrogen, phosphate, or water is not limiting to the plants. Higher temperatures will also increase photosynthetic rates and should promote the upward (elevation) spread as well as the more northward and southward (from the tropics) extension of temperate and tropical species and communities with a simultaneous narrowing of the ranges of cold-tolerant species.

What does this portend for the distribution of the woody legume savannas and woodlands of western and southwestern North America and throughout the world? The potential effects of these anthropogenic changes are difficult to envision over the relatively short term because of the age or potential age that the woody legumes and their communities can attain. Besides, experimental controls are lacking, and the last time high levels of atmospheric CO_2 and temperature occurred, there were no humans on the Earth. High atmospheric temperatures at the surface of the Earth imply that temperate communities and species will migrate farther north or south and to higher elevations. Those species that are cold adapted and cold tolerant, already present in northern and southern regions and higher elevations, will relocate, moving further north or south and up in elevation until they cannot move further and then they will be extirpated.

Prior suggestions involve relatively long-term changes. What is expected regarding the short term? There are expected changes in water, nitrogen, and other nutrients. Dramatic shifts in weather patterns are expected, including more severe droughts and shifts in rainfall distribution, but the effects of these changes on woody legume establishment, encroachment, growth, and community structure are all unknown. Longer-term effects of these potential changes are also speculative at this time. Supplementary studies are needed to clarify the longer-term effects of climate modifications, including rainfall amounts and distribution in addition to water use and storage. These climatic and atmospheric changes will no doubt cause modifications in the overall water cycle, as well as the carbon and nitrogen cycle and their storage, use, and metabolism. As the use and cycling of these resources change, the plant and animal species in the current grasslands, woody legume savannas, and woodland communities will also change. As physical and chemical changes occur, changes in the vertebrate, invertebrate, plant, bacteria, and fungi populations and diversity will follow. These community modifications will be beneficial to some species and detrimental to others. But, what happens to which ones is unclear.

What kind of experiments would help predict what would happen with encroachment and succession of these woody legume species and communities in the future? Measurement of gas-exchange rates at various levels of temperature, light, CO_2, and other resources for the various species of woody legumes and their associated species would be very helpful. The woody legumes should have high rates of gas exchange at high light levels, which is characteristic of early successional species. Species other than the woody legumes that are present in these communities should also be examined, especially understory species and low-density species. These are the species that could be community replacements and future community dominants in the high temperature, high CO_2 environment of the future. Associated or understory species should have lower gas exchange rates at high light levels and relatively high rates at low light levels, which is expected for late successional species. Carbon uptake and possibly growth rates or relative growth rates of these species should be examined at a variety of soil moisture levels, light levels, and elevated CO_2, nitrogen, and temperature levels. Potential woody legume replacement species from dry, more southern, or northern environments (depending on community location) should be examined and evaluated as well.

In the future, soil organisms and soil processes should not be ignored, especially nitrogen fixation and the organisms associated with the woody legume roots. Soil microbial communities will undoubtedly change as atmospheric CO_2 and temperature levels increase. Other atmospheric inputs will change as well, including nitrogen type and amount and other airborne chemicals. All these factors taken together will probably modify soil biotic communities and alter soil biotic and abiotic reactions. One specific reaction of concern would be nitrogen fixation, but there are others including the various nitrogen cycle conversion reactions. These potential changes will probably alter surface plant and animal communities, but how is unknown at this time.

There is still considerable uncertainty with the predictions of global and regional long-term climate changes. However, increases in extreme events are expected and may be very important in the future, but their frequency and intensity are unclear. Rainfall amounts, distribution, duration, and locations as well as water movement and flow seem certain to change in the future, but direction and amount of these changes, and the effects of altered climate, are further unknowns. Similar things appear true for the other abiotic components of this system. As the abiotic components change, the biotic components will change as well.

These woody legume communities are changing over the short term, but their future composition and structure is difficult to predict. They will continue to change, and they will have a different species composition in the world of the future. It appears that these woody legumes are encroaching into various warm season grasslands because of changing environmental conditions, which would include anthropogenic forces. The woody legume communities next appear to undergo successional changes leading to new communities without the woody legumes. Some of these changes are fairly well known, but some of the new communities will have relatively unknown structure and composition. These legume and other communities are biologically, ecologically, and economically quite dynamic and very important.

The understanding gained from past research work and what has been presented here in this volume is the foundation for future work and future understanding of these communities. Envisaging the future composition and structure of these woody legume communities or the communities resulting from the successional changes is an exciting and important venture, and we seem to be approaching the correct perception of them, but there are many unanswered questions.

Index

O.W. Van Auken and J.K. Bush, *Invasion of Woody Legumes*, SpringerBriefs in Ecology 4,   65
DOI 10.1007/978-1-4614-7199-8, © Springer Science+Business Media New York 2013